卓越工程师教育培养计算机类创新系列教材

C++面向对象程序设计

主　编　曲维光　姚望舒

副主编　胡　勇　王东波　田蓓艺　张砚雪

科学出版社

北　京

内 容 简 介

本书面向已学习过 C 语言程序设计的读者，重点介绍 C++面向对象程序设计的内容。全书共 9 章，从 C 到 C++，类和对象的创建，类和对象的使用，运算符重载，继承与派生，多态性与虚函数，文件系统，面向对象程序设计应用举例，以及实验操作。

本书作者多年从事 C++面向对象程序设计的教学工作，并且具有利用 C++进行大型软件设计开发的经验。在注重课程体系完整性的同时，更加关注教学内容的实用性。内容由浅入深，调理清晰，语言流畅；例题丰富，紧扣知识点，具有实用性；基于 Visual C++ 6.0，并同时兼顾 Visual Studio.net 和 GNU C++，全部实例程序可以运行；注重编程方法的讲解，强调对编程素养的培养；每章都有偏重实践能力培养的习题，便于学生评估学习效果；附录提供 4 套模拟试卷，方便学生温课备考或自我检查学习效果。

本书可作为高等学校计算机及其相关专业"C++面向对象程序设计"课程的教材，也可作为从事计算机软件开发人员的参考用书。

图书在版编目（CIP）数据

C++面向对象程序设计 / 曲维光，姚望舒主编. —北京：科学出版社，2016.12
卓越工程师教育培养计算机类创新系列教材
ISBN 978-7-03-049035-3

Ⅰ. ①C⋯ Ⅱ. ①曲⋯ ②姚⋯ Ⅲ. ①C 语言－程序设计－教材 Ⅳ. ①TP312

中国版本图书馆 CIP 数据核字(2016)第 141881 号

责任编辑：于海云 / 责任校对：郭瑞芝
责任印制：赵 博 / 封面设计：迷底书装

科学出版社 出版
北京东黄城根北街 16 号
邮政编码：100717
http://www.sciencep.com

北京科印技术咨询服务有限公司数码印刷分部印刷
科学出版社发行 各地新华书店经销

*

2016 年 12 月第 一 版　开本：787×1092　1/16
2024 年 12 月第六次印刷　印张：19 1/2
字数：462 000
定价：65.00 元
（如有印装质量问题，我社负责调换）

前　　言

C++面向对象程序设计语言是从 C 语言发展而来的一种高级语言，是目前广泛使用的一种程序设计语言。目前大多数 C++教材都包含许多 C 语言的内容，这对于已经学过 C 语言和结构化程序设计方法的读者来说显得篇幅过长，不便于课堂教学。

本书从面向对象程序设计方法展开 C++的内容，适合于对已经学过 C 语言程序设计的学生进行 C++面向对象程序设计课程的教学，也适合已经有 C 语言程序设计基础，希望尽快学习 C++面向对象程序设计的读者阅读。

本书共 9 章。第 1 章从 C 到 C++，主要介绍 C++与 C 语言的区别与联系，C 语言重要知识点回顾及 C++的扩展功能，并介绍输入和输出流的基本知识及文件处理和异常基本知识，便于学生通过使用而熟练掌握。第 2 章类和对象的创建，主要介绍从结构体到类，类声明和对象定义，构造函数和析构函数，对象的复制与赋值，构造函数和析构函数调用的顺序。第 3 章类和对象的使用，主要介绍动态对象的创建与释放，对象数组，建立数组类，类的组合成员，静态成员，友元。第 4 章运算符重载，主要介绍运算符重载的概念方法，运算符重载规则，单目和双目运算符重载，流插入运算符和流提取运算符的重载，不同类型数据间的转换，综合应用实例。第 5 章继承与派生，主要介绍继承与派生的概念，定义基类和派生类，基类成员在派生类中的可访问性，派生类的构造函数和析构函数，多重继承与虚基类，基类与派生类的转换，综合应用例题。第 6 章多态性与虚函数，主要介绍多态性的概念，虚函数，纯虚函数与抽象类，综合应用例题。第 7 章文件系统，主要介绍文件的基本概念，文件的打开和关闭，文件定位函数，成批文件的处理，文件操作程序举例。第 8 章面向对象程序设计应用举例，通过一个较大型问题的解决过程来展示系统分析，系统设计，系统实现，系统运行结果。第 9 章介绍各种实验操作。

本书的主要特点：

（1）选取与知识点内容契合的、贴近实际应用的例题进行讲解，读者可以通过例题更加容易理解知识点；所有例题均可运行，并可以适度扩展，用于实际问题求解。

（2）打破原有教学内容的顺序，将文件系统和异常处理放到前面讲授，从而可以让读者熟练掌握文件处理和异常处理等程序设计中重要的内容。

（3）每章都配有注重实践能力培养的课后习题，便于读者检验学习的效果，使学生不仅能看得懂程序，也能逐步学会编写较大规模的程序。

（4）书后提供 4 套笔试模拟试卷，利于学生温课备考。

本书第 1、4、8 章由姚望舒编写，第 2、3 章由曲维光和田蓓艺编写，第 5、6 章由胡勇和张砚雪编写，第 7 章由王东波和曲维光编写，第 9 章为实验，由曲维光、姚望舒、胡勇和王东波编写。全书由曲维光和姚望舒担任主编，并负责统稿、修改和定稿。

由于作者水平有限，书中难免存在不妥之处，敬请读者批评指正。

编者 E-mail: wgqu_njnu@163.com

编　者
2016 年 4 月

目 录

前言

第1章 从C到C++ 1
1.1 C++与C语言的区别与联系 1
1.2 C语言重要知识回顾及C++的扩展 2
1.2.1 常量 2
1.2.2 指针与引用 3
1.2.3 名字空间 6
1.2.4 C字符串和string类 8
1.2.5 函数 13
1.3 输入/输出流的基本知识 26
1.3.1 标准I/O流 27
1.3.2 文件流 27
1.3.3 流状态 31
1.3.4 文件操作实例 33
1.4 异常基本知识 34
练习 36

第2章 类和对象的创建 38
2.1 从结构体到类 38
2.2 类声明和对象定义 41
2.2.1 类的数据成员和成员函数 42
2.2.2 访问权限 43
2.3 构造函数和析构函数 44
2.3.1 构造函数 44
2.3.2 默认构造函数 47
2.3.3 构造函数的重载 48
2.3.4 参数初始化列表 54
2.3.5 析构函数 58
2.4 对象的复制与赋值 59
2.4.1 复制构造函数 59
2.4.2 深复制和复制构造函数 65
2.4.3 对象的赋值 67
2.5 构造函数和析构函数调用的顺序 71
练习 73

第3章 类和对象的使用 ... 74

- 3.1 动态对象的创建与释放 ... 74
- 3.2 对象数组 ... 77
- 3.3 建立数组类 ... 80
- 3.4 类的组合成员 ... 84
- 3.5 静态成员 ... 89
  - 3.5.1 静态数据成员 ... 91
  - 3.5.2 静态成员函数 ... 96
- 3.6 友元 ... 98
  - 3.6.1 普通函数作为友元函数 ... 98
  - 3.6.2 成员函数作为友元函数 ... 102
  - 3.6.3 友元类 ... 107
- 3.7 综合应用例题 ... 111
- 练习 ... 115

第4章 运算符重载 ... 116

- 4.1 运算符重载的定义 ... 116
- 4.2 运算符重载的方法 ... 117
- 4.3 运算符重载规则 ... 118
- 4.4 单目运算符重载 ... 119
- 4.5 双目运算符重载 ... 123
  - 4.5.1 下标运算符重载 ... 123
  - 4.5.2 赋值运算符重载 ... 126
  - 4.5.3 加法运算符重载 ... 128
- 4.6 流插入运算符和流提取运算符的重载 ... 129
- 4.7 不同类型数据间的转换 ... 132
  - 4.7.1 基本数据类型到类类型的转换 ... 132
  - 4.7.2 类类型到基本类型的转换 ... 134
- 4.8 综合应用实例 ... 135
- 练习 ... 146

第5章 继承与派生 ... 148

- 5.1 继承与派生的概念 ... 148
- 5.2 定义基类和派生类 ... 149
  - 5.2.1 定义基类 ... 149
  - 5.2.2 定义派生类 ... 152
- 5.3 基类成员在派生类中的可访问性 ... 158
  - 5.3.1 公用继承 ... 158
  - 5.3.2 私有继承 ... 162
  - 5.3.3 保护成员和保护继承 ... 166
  - 5.3.4 多级派生时的访问属性 ... 169

5.4 派生类的构造函数和析构函数 ·············· 172
5.5 多重继承与虚基类 ························ 177
    5.5.1 多重继承 ····························· 177
    5.5.2 虚基类 ······························· 181
5.6 基类与派生类的转换 ····················· 183
5.7 综合应用例题 ······························ 185
练习 ··· 189

第6章 多态性与虚函数 ························ 194
6.1 多态性的概念 ······························ 194
6.2 虚函数 ······································· 196
    6.2.1 虚函数的作用 ······················ 196
    6.2.2 虚析构函数 ························· 197
6.3 纯虚函数与抽象类 ························ 200
    6.3.1 纯虚函数 ··························· 200
    6.3.2 抽象类 ······························ 200
    6.3.3 抽象类应用实例 ··················· 201
6.4 综合应用例题 ······························ 205
练习 ··· 212

第7章 文件系统 ··································· 214
7.1 文件的基本概念 ··························· 214
    7.1.1 文本文件和二进制文件 ·········· 215
    7.1.2 标准文件 ··························· 216
    7.1.3 缓冲型文件和非缓冲型文件 ···· 216
7.2 文件的打开和关闭 ························ 217
    7.2.1 字符级读写 ························ 221
    7.2.2 字符串级读写 ····················· 222
    7.2.3 格式化读写 ························ 223
    7.2.4 二进制数据读写 ·················· 225
7.3 文件定位函数 ······························ 227
7.4 成批文件的处理 ··························· 229
    7.4.1 文件名骨架的设计 ··············· 229
    7.4.2 库函数_findfirst和_findnext ···· 229
    7.4.3 批处理文件函数构建 ············ 230
7.5 文件操作程序举例 ························ 233
练习 ··· 235

第8章 面向对象程序设计应用举例 ······· 236
8.1 问题提出 ···································· 236
8.2 系统设计 ···································· 237
8.3 系统实现 ···································· 238

  8.4 系统运行结果 255
  练习 258

第9章 实验 259
  实验1 指针、引用和函数重载 259
  实验2 文件操作 260
  实验3 类的建立和使用 260
  实验4 运算符重载 261
  实验5 继承与派生 261
  实验6 多态性和虚函数 263
  实验7 文件系统的操作 270

参考文献 271

附录 272
  模拟试卷1 272
  模拟试卷2 277
  模拟试卷3 283
  模拟试卷4 293

第 1 章　从 C 到 C++

C++是在 C 语言基础上发展演变而来的一种面向对象的编程语言，C++继承了 C 语言的大部分特征，同时也对 C 语言进行了大量的扩展，是一种全新的程序设计语言。换句话说，C 语言是 C++的一个子集。

- 知识要点

  掌握 C++对 C 语言的一些扩展知识：引用、函数重载；

  初步掌握文件流的使用、异常的基本概念及使用。

- 探索思考

  函数的参数传递机制。

- 预备知识

  复习 C 语言的指针、数组、字符串等知识；

  复习函数及函数参数传递；

  复习文件的概念及操作。

1.1　C++与 C 语言的区别与联系

C 语言是一种通用的程序设计语言，适合大部分系统程序设计任务。C 语言程序可以在很多系统上运行，并且语法简洁、灵活，数据结构丰富，兼有高级语言和汇编语言的优点，能直接访问物理地址，与汇编语言相比，具有良好的可读性和可移植性，程序执行效率高。C 语言应用广泛，甚至连 UNIX 操作系统也是用 C 语言编写的。但是，C 语言是一种面向过程(Procedure Oriented)的编程语言，随着软件规模的日益扩大，软件功能越来越复杂，C 语言已经难以适用于开发大型程序的要求，其缺点也日益显露。例如，C 语言的类型检查机制较弱，缺乏类型描述机制，对代码重用的支持不够，没有提供对类的支持，无法满足面向对象(Object Oriented)软件开发方法的要求。

针对 C 语言存在的问题，1980 年，贝尔实验室的 Bjarne Stroustrup 创建了一种以 C 语言为基础，支持面向对象的通用程序设计语言——带类的 C 语言，1983 年正式更名为 C++(C plus plus)，经过后来的不断发展完善，形成了目前的 C++。C++在 C 语言的基础上，对 C 语言进行了大量扩展，提出了很多超越 C 语言的功能，其中最重要的功能之一是引入了类的概念，使得 C++可以用于面向对象的软件开发，同时，提供了更便利的代码重用功能。尽管它所提出的覆盖和超越 C 语言的功能变得越来越重要，但 C++语言对 C 语言还是完全兼容的。

C++提出之后的几年，C++的使用出现了爆发式增长。C++的标准化成为一件无法避免的工作。1989 年 12 月，在惠普公司(Hewlett-Packard)的建议下，美国国家标准局(American National Standards Institute, ANSI)成立了 ANSI X3J16 委员会，专门负责制定 C++标准。之后不久，国际标准化组织(International Organization for Standardization, ISO)也成立了自己的 C++标准委员会，负责 C++标准的制定。1991 年 7 月，ANSI 的 C++标准委员会和 ISO 的 C++标准委员会共同合作进行 C++标准化工作，经过委员会的不懈努力，ISO C++标准(ISO/IEC 14882)在 1998 年获得批准。2003 年，委员会制定并发布了 C++标准的第 2 版。2011 年 8 月 12 日，国际标准化组织(ISO)和国际电工委员会(IEC)旗下的 C++标准委员会(ISO/IEC JTC1/SC22/WG21)正式发布 C++11 标准，并于 9 月出版该标准。C++11 标准是 C++98 标准发布以来第一次重大修正。

1.2 C 语言重要知识回顾及 C++的扩展

本节回顾 C 语言的一些重要知识，同时介绍 C++中一些改进和扩展的知识内容。

1.2.1 常量

C 语言中定义的符号常量实际上是一种宏定义，编译器在对程序的预处理过程中完成所有符号常量标识的替换。C 语言中的符号常量定义存在以下不足：

(1)编译器对符号常量进行预处理时，不做任何类型检查，只完成符号常量值的替换。
(2)进行替换时，不做任何合法性检查，这样容易导致编译错误。例如：

```
#define  SUM  3+5
int i_value = 5, i_r=6;
int i_multi = i_value*SUM*i_r;
```

当编译器对上面代码进行预处理时，第 3 行代码会被处理成如下形式：

```
int i_multi = i_value*3+5*i_r;
```

第 3 行语句编译能通过，也能运行，但结果显然不是编程者想要的结果。这类错误的查找会消耗编程者的大量时间。

再如：

```
#define  LEN  3;
int  i_arry[LEN];
```

当编译器对上面代码进行预处理时，第 2 行代码会被处理成如下形式：

```
int i_arry[3;];
```

上面语句显然存在语法错误，但这种错误并不是在预处理时能发现的，而是在编译阶段才能发现。用 Visual Studio 2010 编译该语句时，会指出第 2 行语句存在错误，并不能指出错误是由第 1 行语句引起的，这样增加了查找错误的难度。

C++中提供了一种新的常量定义形式。C++使用关键词 const 定义常量。在定义常量时，const 关键词可以位于类型之前，也可以位于类型与常量标识符之间。例如：

```
const  double  cd_pi = 3.1415;      //定义了 double 常量 cd_pi
```

或者

```
double  const  cd_pi = 3.1415;      //定义了 const 常量 cd_pi
```

上面两种定义常量的形式是等价的，但为了便于阅读，常采用第 1 种形式定义常量。上面的定义中，常量名的前缀 cd 中 c 表示常量，是 const 的首字母，d 表示**常量的类型为双精度实型**，是 double 的首字母。

C++常量的特点：

(1)在程序中，不允许对常量进行修改，即不允许常量作为左值使用。例如：

```
cd_pi = 5.14;        //错误：因为 cd_pi 是常量，不允许作为左值使用。
```

(2)常量必须在定义时进行初始化。因为常量在程序中不能作为左值使用，除了在定义时对其进行初始化之外，没有任何合适的方式初始化该常量。

C++使用 const 定义常量能给编程带来的好处：

(1)C++中，常量是一个具有类型的量，编译器对其进行类型检查和合法性检查，比 C 语言中的符号常量更加安全。例如：

```
const int ci_sum = 3+5;
int i_value = 5, i_r=6;
int i_multi = i_value*ci_sum*i_r;
```

编译器编译上面第 3 行语句时，ci_sum 用计算结果 8 参与运算，因此不会出现 C 语言中符号常量的问题。

(2)在需要改变一个常量值时，能做到"一改全改"。例如，在程序中多处计算图形的面积或体积，如果计算图形面积或体积时π的值用字面常数表示，则需要调整面积或体积的计算精度时，必须对程序中每一个π值的字面常数进行修改。如果用常量 PI 来表示π的值，则只要修改常量 PI 的值，再重新编译程序，就修改了程序中所有π值的精度。

1.2.2 指针与引用

指针是 C++中非常重要的部分。指针使得 C++具有在程序运行时获取变量或对象的地址，并且通过地址操纵数据的能力。指针不但在过程化程序设计中必不可少，而且在面向对象程序设计中也不可或缺。指针对于成功地进行 C++编程至关重要。指针功能强大，能有效地提高程序的性能，但指针又是最危险的，非常容易出错。学习指针时，必须清楚地认识到指针这种双刃剑的作用。

1. 指针

指针变量的定义形式如下：

```
int *ip_pointer;        //定义一个可以指向整型变量的指针变量
char *cp_pointer;       //定义一个可以指向字符变量的指针变量
double *dp_pointer;     //定义一个可以指向 double 变量的指针变量
```

指针变量的定义中，指针变量标识符前面的类型并不是指针变量的类型，而是表示指针变量所能指向的变量类型，也称为指针变量的基类型。在不引起混淆的情况下，本书后面把指针变量称为指针。

指针定义中的*只能修饰紧随其后的变量，如果同一行定义多个指针，则需要每个指针前面都带有*。例如：

```
int *ip_pointer1, *ip_pointer2;     //定义两个可以指向整型变量的指针
```

学习指针时，要理解*号在不同位置和环境下，所表示的含义不同。例如：

```
int *ip_pointer;                //表示 ip_pointer 是一个指针
int i_value1 = 5;
ip_pointer = &i_value1;
```

```
                int  i_value2 = *ip_pointer+3;        //取 ip_pointer 所指向变量的值。
                int  i_value3 = i_value2*i_value1;    //乘法运算符
```

特别提醒： 指针在使用之前必须初始化，使用没有初始化的指针可能会产生无法预料的结果。

1) 指针运算

指针值表示一个内存地址，其内部表示为整数，**指针变量所占的空间大小总是等同于一个整型变量的大小，但指针不是整型数，不遵循整型数的操作规则**。例如，两个指针相乘就没有意义。

虽然指针不遵循整型数的操作规则，但是当指针指向数组时，还是具有如下操作功能：

(1) 当指针指向数组时，指针加上一个正整数 n 表示指针从当前元素位置向后移动 n 个元素；指针减去一个正整数 m 表示指针从当前元素位置向前移动 m 个元素。

(2) 当两个指针变量指向同一个数组时，两个指针之间可以进行比较运算，其结果表示两个指针的相对位置。

(3) 当两个指针变量指向同一个数组时，两个指针之间可以进行减法运算，其结果表示两个指针变量的位置相距多少个数组元素。

特别提醒：

(1) 指针在数组区间的移动是以数组元素类型为单位移动，而不是以字节为单位移动。

(2) 指针加减整数操作时，要检查指针移动的目标位置是否超出了数组的范围，以避免数组越界问题。数组越界是一个严重的问题，也是一个容易出现的问题，所以在程序设计时应该避免该问题的出现。例如：

```
        int i_arry[10] = {1,2,3,4,5,6,7,8,9,10};
        int* pi = i_arry;          //pi 指向第 0 号元素
        pi += 3;                   //pi 指向第 3 号元素，后移 3 个元素
        pi -= 2;                   //pi 指向第 1 号元素，前移 2 个元素
        pi += 9;                   //错：pi 已经指向数组区域之外
        int *pi2 = i_arry+3;       //pi2 指向数组的第 3 号元素
        pi2>pi   //表示了 pi2 和 pi 之间的相对位置，如果 pi2 在 pi 后面，则表达式为 1，否则为 0.
        int i_num = pi2-pi;        //i_num 表示 pi2 和 pi 直接相距的元素个数
```

上面代码中，pi += 3 中，虽然 pi 只后移了 3 个元素，但是，由于每个整型元素在内存中占有 4 字节，pi 移动的具体地址字节数是 12 字节。pi += 9 中，原来 pi 指向数组的第 1 个元素，当 pi 向后移动 9 个元素时，pi 已经指向数组区域之外的地址空间，此时，虽然程序不会引起编译错误，但 pi 所指向的数据没有具体意义，使得后面程序中 pi 所参与的计算结果都是错误的，有时甚至会出现运行异常。pi2>pi 比较了两个指针变量的值，其结果表示两个指针的相对位置。结果大于 0，表示 pi2 在 pi 后面；结果等于 0，表示 pi2 和 pi 指向了同一个元素；结果小于 0，则表示 pi2 在 pi 前面。

2) 指针的限定

当 const 关键词与指针结合时，根据 const 所修饰的对象不同，产生了指针的两种不同的常量性。一种是指针常量(constant pointer)，表示指针本身是常量；另一种是常量指针(pointer to constant)，表示指针所指向的对象是常量。

指针常量定义中，const 修饰指针本身，表示指针的值不能被修改，也就是在程序中不能修改指针的指向。指针常量具有常量的特点，在定义时必须初始化，不能作为赋值表达式左值。

常量指针定义中，const 修饰指针所指向的对象，表示其指向的对象是常量。注意理解指针所指向的对象是常量的概念，在此处，指针所指向的对象是常量并非说对象本身一定是常量对象，而是指通过指针间接访问对象时，对象具有常量性。

下面语句演示了指针常量和常量指针的定义和使用。

```
const int ci = 1;          //定义常量 ci;
int a = 2;
int b = 3;
const int *cpi1 = &ci;     //常量指针，const 修饰所指向的实体对象，该实体对象为常量
const int *cpi2 = &a;      //常量指针，所指向的实体对象为变量
int const *cpi3;           //常量指针，const 位于基类型与*之间
int *const pci1 = &b;      //指针常量，const 修饰指针本身，指针指向不可变
const int *const cpci=&ci; //指向常量的指针常量，常量指针常量
*cpi1 = 3;                 //错：常量指针不能修改所指向的实体对象，*cpi1 只能作右值
*cpi2 = 3;                 //错：理由同上，虽然 a 是变量，但也不能通过*cpi2 修改
cpi3 = &b;                 //对：常量指针本身可为左值
*pci1 = 4;                 //对：指针常量可以修改所指向的实体对象，*pci1 可以作左值
pci1 = &a;                 //错：指针常量不可以改变其的指向
cpci = &a;                 //错：指针常量不可以改变其的指向
*cpci = 5;                 //错：指向常量的指针常量不可作为左值
int *const pci2=&ci;       //错：不可以将非常量指针指向常量
```

上面程序语句中，*cpi1 = 3 和*cpi2 = 3 是错误的，说明常量指针只能作为右值使用，其所指向的对象既可以是常量，也可以是变量，但通过常量指针间接访问对象时，都认为该对象是常量。int *const pci2=&ci 说明在 C++中，只有常量指针才可以指向常量对象。因为如果允许非常量指针指向常量对象，则通过指针的间接访问就可以修改常量对象，这与常量对象的常量性质相矛盾。

指针的限定常用于函数参数传递。

2. 引用

引用(reference)是一个具体对象的别名(alias)。当定义引用时，必须使用一个具体类型的实体来初始化引用，从而在实体与引用之间建立关联，并且这种关联关系是不可以改变的。引用拥有与其关联实体相同的访问待遇。

引用的定义形式如下：

```
int a=30;
int b;
int &ra=a;        //定义引用 ra,其引用实体为 a
int &rb=b;        //定义引用 rb,其引用实体为 b
```

引用就是其关联实体的另外一个名字，对引用的操作就是对其关联实体的操作，修改引用的值就是修改实体的值，引用的地址就是实体的地址。引用与指针不同，引用只是关联实体的一个名字，定义引用并不产生新的实体，并且引用在定义时就确定了其关联的实体对象，这种关联关系是不可改变的，直到引用自身灭亡。

使用引用必须遵循的原则：

(1)引用的类型与其引用的实体类型必须严格一致，否则无法编译通过。

```
float x;
float &rx1=x;        //对:引用的类型与其引用实体类型一致
int &rx2=x;          //错:引用的类型与其引用实体类型不一致
```

(2) 引用在定义时就必须初始化,确定引用与实体之间的关联关系,并且这种关联关系是一直不变的。对引用的操作等价于对其引用实体的操作。

```
int a,b=5;
int &ra=a;           //对
int &rb;             //错:引用在定义时必须初始化
ra=b;                //对
```

上面 ra=b 并不是将引用 ra 的引用实体改变为变量 b,而是将 b 的值赋值给引用 ra 所引用的实体对象 a,使得变量 a 的值为 5。

(3) 给引用初始化的必须是一个内存实体,可以是变量,也可以是常量。

```
int a=5;
int &ra=a;           //对:ra 引用的实体为变量 a
int &rb=int;         //错:int 是类型,不是一个内存实体
```

上面 int &rb=int 用 int 类型初始化引用 rb 是错误的,因为 int 是类型,不是一个内存实体。同样,也不可以定义数组的引用。

从物理实现上理解,引用就是一个隐性指针。引用对实体的访问看似直接访问,实际是指针的间接访问,只是这种转换由编译器在幕后完成。例如:

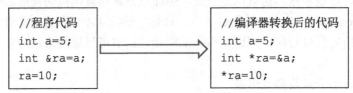

上面代码中 int &ra=a 定义了引用 ra,并初始化引用对象为 a,实际上编译器会将 ra 转换成指针*ra。ra = 10 代码实际是通过指针*ra 间接访问变量 a,完成对变量 a 的赋值操作。

虽然引用仍是以指针方式实现对对象的访问,但它屏蔽了指针实现的地址操作,以直接访问形式完成对象的间接访问,使间接访问更加安全。

引用也可以限定,例如:

```
int a=5;
const int &ra=a;     //定义常引用, ra 不可为左值
a=10;                //对
ra=10;               //错:常引用不可为左值
```

上面代码 const int &ra=a 定义的 ra 具有常量性质,不允许通过 ra 修改实体对象 a,但并不要求 a 必须是常量,这和常量指针相似。**带限定的引用常用于函数参数传递**,引用多用于高级编程,如上层软件开发;而指针多用于低级编程,如硬件驱动等。

1.2.3 名字空间

软件开发方法中,最重要的问题就是程序的合理组织。现代软件开发中,一般需要将软件分成多个可以独立开发的模块,不同模块分配给不同软件设计人员进行开发,或者几个模

块由软件设计人员在不同阶段完成开发。这样分模块多人共同完成软件开发的方式提高了软件开发的效率，但也必然带来一个问题：不同模块之间的标识符(名称)冲突问题。

C++中引入名字空间的概念将标识符冲突问题限定在一个特定的作用域内。它规定，一个标识符只有在使用的域中明确声明其使用的"空间名"，才能在域中默认地使用该标识符。例如：

假如某学校 1 班有一名学生叫"李三"，2 班也有一名学生叫"李三"。这时，对于该学校来说，名字"李三"代表两个不同学生，就存在名称冲突。如何区分两个称为"李三"的学生？这时通过在每个学生的名字前面加上班级名称，如"1 班李三"和"2 班李三"。那么这个"1 班"和"2 班"就是所谓的名字空间，从而解决了名称冲突问题。

C++名字空间的定义形式：

```
namespace  标识符
{
    //名称声明或定义
}
```

例如，下面代码定义了名字空间 myspace，其包含全局整型变量 gicount、函数 print_info 等。

```
namespace  myspace
{
    int gicount;                          //全局计算器
    void print_info(char *pstring);       //输出信息字符串
}
```

名字空间有三种使用方式：

(1) 在使用名字空间中的变量或函数时，指明该变量或函数的名字空间作用域。例如：

```
int  count = myspace::gicount;
```

其中，运算符(::)是域作用运算符，用于指定标识符的作用域，如上面指明 gicount 标识符是名字空间 myspace 作用域的一个全局变量。

这种方式必须在每一个使用名字空间的变量或函数之处显式指定名字空间作用域，当使用之处很多时，增加了代码编写工作量，优点是完全将标识符冲突问题限定在名字空间作用域内。

(2) 在程序中显式声明名字空间的某个或某几个变量或函数，使得在使用这些变量或函数时，就如同使用相同作用域的标识符一样。例如：

```
using myspace::gicount;     //显式声明名字空间 myspace 标识符 gicount
int count = gicount;        //此时可以将 gicount 作为相同作用域的标识符使用
```

当程序需要经常使用到其他名字空间的某个或某几个标识符时，可以将这些标识符进行显式声明，程序就可以将其看成是相同作用域内的标识符。

这种方式将声明的标识符名字冲突扩展到两个名字空间中，所以在使用时需要仔细处理，否则会引起新的名字冲突，好处是大部分名字冲突问题仍限定在名字空间作用域内。

(3) 在程序中显式声明整个名字空间，导致名字冲突问题扩展到了两个名字空间作用域内，名字空间对标识符冲突失去了作用。例如：

```
    namespace myspace2
    {
        using namespace myspace;           //声明名字空间myspace
        int ginum;                         //全局变量
        void myprint (char *pstring);      //输出信息字符串
    }
```

在名字空间 myspace2 中，用语句"using namespace myspace;"显式声明了名字空间 myspace，从而 myspace 中的所有标识符与 myspace2 中的标识符具有相同作用域，此时，myspace2 中不能再定义与 myspace 中相同的标识符，否则就会引起标识符冲突问题。

std 是 C++标准库的名字空间。在编写程序时，因为经常需要用到标准库的类，一般就直接在程序中声明标准库名字空间。例如：

```
/****************************************************************
    File name      :  f0101.cpp
    Description :  标准名字空间的使用
****************************************************************/
#include <iostream>
using namespace std;
int main()
{
    cout << "这是一个名字空间演示程序\n";
    return 0;
}
//==============================================
```

程序的输出结果为：

  这是一个名字空间演示程序

上面程序中，用#include <iostream>方式包含了 C++标准库中标准输入输出类库头文件。用 using namespace std 声明标准库的名字空间 std，所以程序中可以直接使用标准输出流对象 cout 将数据输出到标准设备(屏幕)，不需要再指定 cout 的名字空间作用域。如果没有对标准库名字空间的声明，则必须为下面语句形式，显式指定 cout 的名字空间作用域。

  std::cout<<"这是一个名字空间演示程序\n";

因为程序中对标准库名字空间的声明，则程序中不能再定义与 std 名字空间中标识符名称相同的标识符。

1.2.4 C 字符串和 string 类

C++中有两种字符串：一种是从 C 语言继承下来的 C 字符串；另一种是由 C++标准模板库(Standard Template Library，STL)提供的 string 类。

1. C 字符串

C 字符串是以一个全 0 位(整数 0)字节作为结束符的字符序列。该全 0 字节既是 8 位的整数 0，又是 ASCII 码为 0 的字符。C 字符串是以字符数组形式存放，字符数组长度必须比字符串长度多一个字节，用于存放字符串结束标记。

C 字符串的赋值、复制、比较、连接等操作都是通过 C 库函数来实现的，常用的字符串操作库函数见表 1-1。

表 1-1 常用的字符串操作库函数

函数原型	函数功能	返回值
char* strcpy(char* str1, char* str2)	复制 str2 指向的字符串到 str1 中	返回 str1
int strcmp(char* str1, char* str2)	比较字符串 str1 和 str2	str1<str2，返回负数 str1==str2，返回 0 str1>str2，返回正数
unsigned int strlen(char* str)	统计字符串 str 中字符个数，不包括终止符'\0'	返回字符个数
char* strcat(char* str1, char* str2)	把字符串 str2 连接到 str1 后面，str1 原来的终止符取消	返回 str1

在 C++中，这些库函数的操作，默认在 cstring 头文件中声明。

由于 C 字符串是以字符数组形式存放，C 字符串的使用容易出现数组越界、内存泄漏等问题，使用时需要特别注意。例如：

```
char* str1;
char* str2 = new char[5];
strcpy(str2, "ugly");
strcpy(str1,str2);           //错:str1 没有空间可存储字符串 str2 的字符
strcpy(str2, "Hello");       //错:str2 空间不够大
str2 = "Hello";              //错:原来的"ugly"空间脱钩，导致内存泄漏
```

上面代码中第 2 行使用 C++的 new 操作符在内存中申请长度为 5 的字符数组，并用指针 str2 指向该内存区域，如图 1-1 所示。

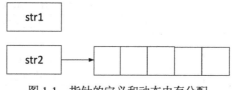

图 1-1 指针的定义和动态内存分配

第 3 行将字符串"ugly"复制到 str2 所指的内存空间中，如图 1-2 所示。

图 1-2 字符串复制

第 4 行希望将 str2 指向的字符串复制到 str1 中，但 str1 没有初始化指向任何有效地址空间，所以复制操作无法实现。第 5 行字符串的长度与 str2 的空间相同，没有空间存放字符串结束标记'\0'，所以 str2 没有足够大的空间存放字符串，字符串复制失败。第 6 行改变 str2 的指向，使得原来使用 new 操作符申请的内存空间与 str2 脱钩，导致该部分内存空间再也无法使用，直到程序退出，由操作系统收回，这就是软件设计中一个非常严重的问题——内存泄漏，如图 1-3 所示。

图 1-3 内存泄漏

C++提供了两个用于动态内存管理的操作符，分别是 new 和 delete，new 用于申请内存空

间，delete 用于释放内存空间。new 和 delete 与 C 语言中的 malloc 函数和 free 函数功能相同，但更安全。关于 new 和 delete 操作符的详细介绍请参考第 3 章对象的动态建立与释放。

2. string 类

string 类是 STL 提供的一种自定义类型，它可以方便地执行 C 字符串所不能直接完成的一切操作，它可以将字符串(string)当作一个类似于整型量的一般类型使用，而不会引起任何问题。可以像对待基本数据类型那样地赋值、复制和比较等，让用户再也不用担心内存是否足够。STL 对 string 类的设计思维就是让它的行为尽可能像基本数据类型一样，不会在操作上引起什么问题。

string 类为字符串设计了相应的操作函数，包括一些运算符重载函数。有关运算符重载函数的相关知识见第 4 章。string 类常用的操作函数见表 1-2。

表 1-2 string 类常用操作函数

操作函数	功　　能
=，assign()	赋值操作函数
swap()	交换两个字符串的内容
+=，append()，push_back()	添加字符
insert()	在参数规定位置插入字符
erase()	删除参数规定位置的字符
clear()	移除全部字符，string 类对象为空
replace()	替换字符
+	字符串拼接
size()，length()	返回字符数量
==，!=，<，>，<=，>=，compare()	字符串的比较功能
empty()	判断字符串是否为空
[]，at()	存取单个字符
>>，getline()	从 stream 流中读取某值(有关流的概念见 1.4)
<<	将某值写入 stream
copy()	将内容复制为一个 C 字符串
c_str()	将内容以 C 字符串形式返回
data()	将内容以字符数组形式返回
substr()	返回某个子串

string 类在 string 头文件中定义，下面程序演示了 string 类的基本使用方法。

```
/******************************************************************
    File name   : f0102.cpp
    Description : 演示 string 类的基本使用方法
******************************************************************/
#include<iostream>
#include<string>
usingnamespace std;

int main()
    {
        string ls_str1;
```

```cpp
    string ls_str2 = "ABCDEFGH";
    string ls_str3 = "abcdefgh";

    ls_str1 = ls_str2;                         //赋值操作
    cout << ls_str1 << endl;
    ls_str1.swap(ls_str3);                     //交换操作
    cout << ls_str1 << endl;
    cout << ls_str3 << endl;
    ls_str1 += 'A';                            //添加字符
    cout << ls_str1 << endl;
    ls_str1.insert(1, "B");                    //插入字符串
    cout << ls_str1 << endl;
    ls_str1.erase(5, 2);                       //删除字符
    cout << ls_str1 << endl;
    cout << ls_str1 + ls_str3 << endl;         //字符串连接
//比较
if (ls_str1 > ls_str3)
    {
        cout <<"ls_str1>ls_str3.\n";
    }
elseif (ls_str1 == ls_str3)
    {
        cout <<"ls_str1 and ls_str3 are equal.\n";
}
else
    {
        cout <<"ls_str1 < ls_str3\n";
    }
//存取、下标和 at()函数
    ls_str1[4] = 'F';
    cout << ls_str1 << endl;
for (unsignedint i = 0; i < ls_str1.length(); i++)
    {
        ls_str1[i] += 1;                       //所有字母的 ASCII 值加 1
    }
    cout << ls_str1 << endl;
    cout << ls_str1.at(4) << endl;
    ls_str1.at(4) = 'H';
    cout << ls_str1 << endl;

    string ls_str4 = "0123456789";
cout << ls_str4 << endl;
    ls_str1 += ls_str4.substr(1, 4);           //提取子串
cout << ls_str1 << endl;

    ls_str1.clear();                           //移除字符串
if (ls_str1.empty())
    {
```

```
            cout <<"ls_str1 is cleared! "<< endl;
        }
    
    return 0;
}
//================================================================
```

程序的输出结果为:

```
ABCDEFGH
abcdefgh
ABCDEFGH
abcedfgA
aBbcdefghA
aBbcdghA
aBbcdghAABCDEFGH
ls_str1>ls_str3
aBbcFghA
bCcdGhiB
0123456789
bCcdGhiB1234
ls_str1 is cleared!
```

程序首先通过 include 包含了 C++中标准输入输出头文件。C++中是通过输入输出流类来实现数据的输入输出。通过#include <string>包含了 string 类的头文件。

使用 using namespace std 声明了 C++中的**标准名空间**。

定义了三个 string 类实体对象 ls_str1、ls_str2、ls_str3。string 类实体对象的定义形式与基本数据类型定义形式一样。如程序中定义了一个没有初始化的 string 类实体对象 ls_str1 和两个带初始化的实体对象 ls_str2、ls_str3。string 类实体对象还有参数化的定义形式，例如：

```
    string str1("ABC");
    string str2(5,'A');
    string str3(5);
```

上面语句中，string 实体对象的定义根据参数将 str1 初始化为"ABC"，将 str2 初始化为包含 5 个字符'A'的字符串，将 str3 初始化为包含 5 个空格的字符串。

第 11 行通过 string 类的成员函数实现两个字符串的交行。

第 16 行通过 insert 函数在字符串中指定位置插入一个字符串。该函数的第 1 个参数表示插入字符串的起始位置，第 2 个参数表示要插入的字符串，可以是一个 C 字符串，也可以是另外一个 string 类实体对象。

第 18 行通过 erase 函数删除字符串中指定位置开始的多个字符。该函数的第 1 个参数表示删除字符的起始位置，也就是删除字符的起始字符索引值，第 2 个参数表示从该位置开始删除多少个字符。

第 35～44 行演示了 string 类的存取操作。string 类两种方式的存取方式：一种是通过下标方式。使用该方式时，要特别注意下标越界问题。如果下标越界，其行为未定义，可能产生无法预知的结果。另一种方式是通过 at()函数来存取单个字符。该方式会检查索引是否有效，如果无效，则会抛出 out_of_range 异常。有关异常基本知识见 1.4 节。

特别提醒：在确定下标边界时，必须使用 string 类的 size 或 length 函数确定最大下标值，不可以使用字符串结束标记（'\0'）确定下标边界，因为字符（'\0'）在 string 中是一个有效字符。另外，如果 string 类实体对象没有初始化就对其进行存取，则会引起异常，此时 string 类实体并没有申请空间用于存储字符，会产生 out_of_range 异常或下标越界问题。例如：

```
string str1;
str1[0] = 'A';
```

上面语句中，首先定义 str1 对象，但没有初始化，则第 2 行的语句通过下标方式修改 str1 的某个字符时，会引起 out_of_range 异常。

第 47 行提取 ls_str4 第 1 号字符开始的 4 个字符，然后附加到 ls_str1 后面。substr 函数用于从 string 类中提取子串，函数的第 1 个参数表示提前子串的起始字符索引值，第 2 个参数表示提取的字符个数。

string 类型还有很多其他操作函数，在此不一一举例介绍，要学会使用 string 类的所有功能，请查阅关于标准模版库教程资料。

1.2.5 函数

函数是 C 语言和 C++程序结构中的基本单位，只是 C++函数在内涵上对 C 语言进行了扩展。C++的函数是一个过程，有些函数可以有输入参数和返回值，如数学函数；有些函数可能没有参数或返回值。C++中函数参数的意义不仅仅是一个类型的变量值，很多时候是指向数据流的一个窗口，或者数据集合以及一组相关的操作，也就是对象。

函数在 C++中非常重要，所以函数使用必须十分规范。在学习和使用函数时，要特别理解函数参数的性质、函数参数的传递规则、函数返回值类型、函数调用时的规范等方面。

1. 函数的定义与调用

1) 函数定义

从用户使用的角度来看，C++函数可以分为两类：标准库函数和用户自定义函数。标准库函数由 C++系统定义，并提供给用户使用。用户自定义函数由用户根据特定任务编写的函数。用户自定义函数的定义形式如下：

```
返回值类型 函数名(<形式参数列表>)
{
    //函数体代码
}
```

其中，返回值类型指调用函数后所得到的函数值的类型，可以是各种数据类型，根据函数功能来确定。函数的返回值通过 return 语句返回给调用函数，例如：

```
return 表达式;
```

如果函数无返回值，则函数返回值类型为 void，此时，函数体中可以省略 return 语句。例如：

```
void print_str(string str)
{
    cout<< str;
```

```
            //return ; 语句可以省略;
        }
```

形式参数列表中的形式参数又称为形参,是函数与其他函数传输数据的通道。在函数定义中,形式参数列表并不是必须的。如果一个函数没有形式参数,则称为无参函数。有形式参数的函数称为有参函数。形式参数列表一般为如下形式:

<类型 1><形参名 1>,<类型 2><形参名 2>,…,<类型 n><形参名 n>

函数定义中,形参列表中的形参只是一个符号,用于告知使用者形参出现的位置需要一个什么类型的数据。

函数体是用于实现函数功能的各种程序语句序列。

特别提醒:

(1)函数调用返回结果值的类型由函数定义时的返回值类型确定,不是由 return 语句中表达式值的类型确定。如果 return 语句的表达式值类型与函数定义的返回值类型不同,则系统会将表达式值转换为函数返回值的类型。

(2)函数调用之前必须先对函数进行声明。

函数声明语句形式如下:

返回值类型函数名(<形式参数列表>);

当函数在调用该函数的函数之前定义,则可以省略单独的函数声明语句。例如:

```
void print_str(string str)
{
    cout<< str;
}
void funA(string str)
{
    //函数代码...
    print_str(str);
}
```

函数 print_str 在函数 funA 之前定义,所以在 funA 函数之前可以省略 print_str 函数的声明。

可以在程序的任何地方对函数进行多次声明,但必须保证所有函数声明是一致的,否则会导致编译错误。

2) 函数调用

函数是通过函数调用语句来发挥作用的。函数调用指定了被调函数的名字以及调用该函数所需要的信息,也就是参数。调用函数时所提供的参数称为实际参数,简称为实参。通常称调用其他函数的函数为主调函数,被调用的函数为被调函数。例如:函数 funA 调用函数 funB,则 funA 称为主调函数,funB 称为被调函数。主调函数和被调函数的关系是相对的,只表示了一次函数调用过程中,两个函数之间的关系。例如:当函数 funC 调用函数 funA,则 funC 为主调函数,而 funA 就称为被调函数。

函数调用的一般语句形式:

函数名(<实参列表>);

其中,实参列表中各参数用逗号分隔,实参可以是常量、变量、表达式,甚至可以是函数的返回值。**注意**:函数调用中的实参与被调函数形参的个数、顺序和类型必须一致,才能正确匹配。下面程序演示了函数定义和调用的基本过程。

```cpp
/*******************************************************************
    File name    :  f0103.cpp
    Description  :  函数调用的基本过程
*******************************************************************/
#include<iostream>
usingnamespace std;

int max(int x, int y)
{
    return (x > y ? x : y);
}

int main()
{
    int x, y, z;
    cout <<"Please input 3 number:\n";
    cin >> x >> y >> z;
    cout <<"The max value: "<< max(x, max(y, z)) << endl;
    return 0;
}
//================================================================
```

程序的输出结果为:

```
Please input 3 number:
4 2 5
The max value: 5
```

程序中定义了一个计算两个整型数的最大值函数 max，该函数有两个整型形式参数，返回两个参数的最大值。

在 main 函数中调用 max(x,max(y,z))是作为输出语句中的表达式使用，其中 max(y,z)函数的返回值又作为外层 max 函数的实参使用。

在程序中，形式参数为 x 和 y，实际参数为 main 函数中的局部变量 x、y、z，拥有具体内存空间的实体变量。形参和实参分别属于两个不同的作用域，形参 x 和 y 的作用域为 max 函数，实参 x、y、z 的作用域为 main 函数，所以参数名相同也不会引起错误。

2. 函数的参数传递

1) 程序运行时内存布局

一个程序需要操作系统将程序的可执行程序文件装载到计算机内存中才能运行。一般而言，操作系统将程序装入内存后，就形成一个随时可以运行的进程空间，该进程空间可分为四个区域，如图1-4所示。

代码区存放程序的执行代码，由函数定义块编译生成。

全局数据区存放程序中的所有全局数据、常量、文字量、静态全局变量和静态局部变量。

堆区存放程序中产生的动态内存，供程序随机申请使用。如

进程空间
代码区 (code area)
全局数据区 (data area)
堆区 (heap area)
栈区 (stack area)

图1-4　运行中的内存布局

果程序中反复申请动态内存,且没有在适当时机释放,则会慢慢将程序的整个堆区占有,导致程序崩溃,这就是常说的内存泄漏。

栈区存放函数数据(即局部数据),其动态反映了程序运行中的函数状态。

2)函数的运行机制

栈是一种数据结构,其工作原理是后进先出,就如同将盘子叠在一起,先放的盘子叠在最下面,后放的盘子叠在上面,下面的盘子只有等到上面的盘子拿走才可以拿出。

C++中函数调用过程就是在栈空间操作的过程。函数调用过程中,C++需要完成以下工作:

(1)建立被调函数的栈空间,栈空间大小由函数定义体中的数据量决定。

(2)保护调用函数的运行状态和返回地址。

(3)传递参数。

(4)将控制权交给被调函数。

(5)函数运行完成后,复制返回值到函数数据块底部。

(6)恢复调用函数的运行状态。

(7)返回调用函数。

例如,程序 f0103 中,main 函数调用 max 函数的过程如图 1-5 所示。

max	形参 y	5
	形参 x	2
	返回 max 地址	
	max 状态保护例程指针	
	返回值	5
max	形参 y	max(y,z)的返回值:5
	形参 x	4
	返回 main 地址	
	main 状态保护例程指针	
	返回值	5
main 函数	x	4
	y	2
	z	5
	参数	
	返回操作系统地址	
	操作系统状态保护例程指针	
	操作系统返回值 0	

图 1-5 函数调用过程

C++中调用一个函数可以看做是一个栈中元素的压栈和退栈操作。压栈之后的函数仍可能调用其他函数,因此,程序运行表现为栈中一系列元素的压栈和退栈操作。最初,操作系统将 main 函数压入栈中,标志程序开始运行,最后 main 函数退栈,标志程序结束运行。

函数参数传递是指在函数调用过程中,形参与实参之间进行数据交互的过程。在函数调用前,形参只是一种标识,在内存中没有存储空间,也没有具体值。函数调用时,系统才在栈区为形参分配内存空间,当函数执行完毕,退出后,系统将收回函数形参在栈区占有的内存空间。C++中,有两种形式的函数参数传递,分别是传值形式和传地址形式。**注意,无论是哪种参数传递形式,函数的参数传递永远是从实参到形参的单向传递。**

3) 传值形式

在传值形式中，实参值被复制到对应形参对象中，作为形参的初始值。函数中对形参的访问、修改等操作都是在形参对象上完成，与实参无关，不会引起实参值的变化，实参和形参是完全不同的内存空间。

```cpp
/******************************************************************
    File name    : f0104.cpp
    Description  : 演示函数参数传递的传值形式中形参和实参的关系
******************************************************************/
#include<iostream>
usingnamespace std;

void my_swap(int, int);      //函数声明
int main()
{
    int x, y;
    cout <<"Please input two number:\n";
    cin >> x >> y;
    cout <<"Before swapping\n";
    cout <<"x="<< x <<",y="<< y << endl;
    my_swap(x, y);
    cout <<"After swapping:\n";
    cout <<"x="<< x <<", y="<< y << endl;
    return 0;
}
voidmy_swap(int a, int b)
{
    int temp=a;
    a=b;
    b=temp;
}
//===================================================================
```

程序的输出结果为：

```
Please input two number:
4 5
Before swapping:
x=4, y=5
After swapping:
x=4, y=5
```

因为 my_swap 函数在 main 函数之后定义，C++规定在调用函数之前必须声明函数，所以程序在第 3 行对 my_swap 函数进行声明。在函数声明语句中，形参只是标识函数各个参数的类型，可以省略形参名。

从输出结果看，函数 my_swap 没有交换实参 x 和 y 的值，没有达到预期效果。在第 11 行主函数调用 my_swap() 函数时，系统在栈区建立 my_swap 函数形参 a 和 b，实参 x 和 y 的值分别复制到形参 a 和 b 中。之后，my_swap 函数对形参的操作与实参没有任何关系。退出

my_swap 函数之后，栈区中的形参 a 和 b 被撤销。my_swap 函数并不能改变实参 x 和 y 的值。具体过程见图 1-6。

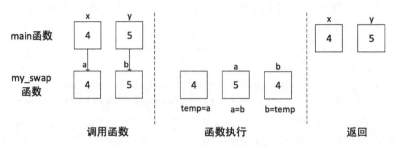

图 1-6　传值形式的函数调用过程

4）传地址形式

从程序 f0104 中可以看出，函数参数的传值形式并不能完成函数所期望的功能，无法在函数中操作实参，即无法操作主调函数的数据。如果被调函数需要操作实参，则需使用另外一种参数传递形式——传地址形式。C++中有两种方式实现传地址形式，分别是指针和引用作为函数参数。C++中传地址形式的参数传递方式将主调函数的数据地址传递给被调函数，使得被调函数拥有了操作主调函数数据的能力。虽然被调函数能通过参数修改主调函数的数据，但函数参数传递仍然是从实参到形参的单向传递，只是此时传递的是实参所指向的数据地址值。

(1) 指针传递

指针作为函数形参时，实参到形参的传递是将主调函数的数据地址传递给形参，从而使得被调函数通过形参具有操作主调函数数据的能力。

```
/****************************************************************
    File name    :  f0105.cpp
    Description  :  演示指针作为函数参数时形参和实参的关系
****************************************************************/
#include<iostream>
usingnamespace std;

void my_swap(int*, int*);        //函数声明
int main()
{
    int x, y;
    cout <<"Please input two number:\n";
    cin >> x >> y;
    cout <<"Before swapping\n";
    cout <<"x="<< x <<",y="<< y << endl;
    my_swap(&x, &y);
    cout <<"After swapping:\n";
    cout <<"x="<< x <<", y="<< y << endl;
    return 0;
}
voidmy_swap(int *px, int *py)
{
```

```
    int temp = *px;
    *px = *py;
    *py = temp;
}
//=================================================================
```

程序的输出结果为：

```
Please input two number:
4 5
Before swapping:
x=4, y=5
After swapping:
x=5, y=4
```

输出结果表明，my_swap 函数修改了 main 函数中 x 和 y 的值，my_swap 函数实现了交换两个整型变量的值。函数调用过程如图 1-7 所示。

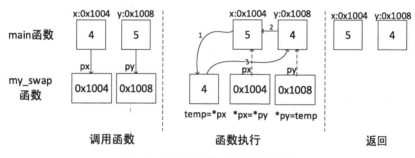

图 1-7 指针为参数的函数调用过程

图 1-7 中 x:0x1004 表示变量 x 的地址，同样，y:0x1008 表示变量 y 的地址。图中虚线表示指针指向变量。main 函数中的 my_swap(&x,&y) 语句将 main 函数中变量 x 和 y 的地址赋值给形参 px 和 py，使得 my_swap 函数拥有了操作 main 函数中变量 x 和 y 的能力。在函数执行过程中，首先将形参 px 所指变量 x 的值保存到 my_swap 函数的局部变量 temp 中，其次将形参 py 所指变量 y 的值赋值给形参 px 所指变量 x，最后将局部变量 temp 的值赋值给形参 py 所指变量 y。my_swap 函数通过实参传递给形参的地址交换了主调函数 main 中两个变量的值。

(2) 引用传递

引用参数的效果与指针参数的效果一样，也是一种传地址形式，使得被调函数拥有操作主调函数数据的能力。引用参数的表现形式与指针完全不同，引用参数的形参为引用，在函数调用时实参为对象名，形式上与传值形式相同。

```
/*****************************************************************
    File name    : f0106.cpp
    Description  : 演示引用作为函数参数时形参和实参的关系
*****************************************************************/
#include<iostream>
usingnamespace std;
void my_swap(int&, int&);        //函数声明
int main()
{
```

```cpp
    int x, y;
    cout <<"Please input two number:\n";
    cin >> x >> y;
    cout <<"Before swapping\n";
    cout <<"x="<< x <<",y="<< y << endl;
    my_swap(x, y);
    cout <<"After swapping:\n";
    cout <<"x="<< x <<", y="<< y << endl;
    return 0;
}
void my_swap(int& x, int& y)
{
    int temp = x;
    x = y;
    y = temp;
}
//================================================================
```

程序的输出结果为:

```
Please input two number:
4 5
Before swapping:
x=4, y=5
After swapping:
x=5, y=4
```

输出结果表明，my_swap 函数修改了 main 函数中变量 x 和 y 的值，交换了两个整数的值。引用传递的参数传递过程与指针传递相似。引用作为函数参数时，在函数的定义中对数据的操作以及函数调用都与传值形式一样，这样避免了指针取值运算和指针误用所带来的问题，使得程序更加安全。

3. 函数重载

当一个软件程序中，调用函数的数量非常多时，其函数命名的难度就开始出现。因为函数都是全局的，彼此不能重名。C++的名字空间机制在一定程度上解决了函数的命名问题，但也存在另外一种情况，很多函数功能相同，只是其处理的数据类型不同，这类函数在 C 语言中必须定义为不同函数。例如：

```cpp
    void my_swap_int(int*,int*);              //交换两个整数的值
    void my_swap_double(double*,double*);     //交换两个double数的值
    void my_swap_str(string*,string*);        //交换两个string对象的值
```

C 语言中，上面三个功能相似的函数必须使用三个不同名字才能区分，这样不仅增加了函数的命名难度，而且因为需要详细记忆各个操作功能的函数名而增加了这类函数的使用难度。C++致力于通过函数重名来达到简化编程的目的。事实上，从这些函数声明中提供的参数信息已经足够区分所描述的操作。C++完全可以同时定义下列函数而不会引起函数意义上的冲突。

```cpp
    void my_swap(int&,int&);                  //交换两个整数的值
```

```
void my_swap (double&,double&);        //交换两个double数的值
void my_swap (string&,string&);        //交换两个string对象的值
```

事实上，C++编译器能够根据函数参数的类型、数量或排列顺序的差异来区分同名函数，这种技术称为**重载技术**，相应的同名函数称为**重载函数**。重载技术不仅化解了一部分函数命名问题，而且在编程逻辑上更加符合人的思维，方便了表达。

判定两个或多个同名函数是否为重载函数的基本原则：
(1) 函数的参数个数不同。
(2) 函数的参数类型不同。
(3) 函数的参数顺序不同。
上面三个条件只要满足一条，C++编译器就能正确匹配特定函数。

特别提醒：函数的返回值类型不能作为区分同名函数的条件，因为在匹配函数时只根据参数来匹配函数，而不关注函数的返回值。例如：

```
void fun(int);             //对
void fun(double);          //对
void fun(int,int);         //对
void fun(int,double);      //对
void fun(double, int);     //对
int  fun(int);             //错：与第一个函数冲突
```

C++按照下列三个步骤的先后顺序查找匹配函数并调用函数：
(1) 寻找一个严格匹配的函数，如果找到了，就调用该函数。
(2) 如果找不到一个严格匹配的函数，则通过相容类型隐式转换寻求一个匹配，如果找到了，就调用该函数。
(3) 通过用户定义的转换寻求一个匹配，如果能找到唯一的一组转换，则调用该函数。

例如，前面重载的对象交换函数my_swap的调用：

```
int a=5,b=4;
my_swap(a,b);              //严格匹配my_swap(int,int);
float x=2.3,y=3.3;
my_swap(x,y);              //通过相容类型的隐式转换寻求到匹配
```

C++允许低类型到高类型的多种转换，如C++即允许int型到long型的隐式转换，又允许int型到double型的隐式转换。如果在同名重载函数匹配时，需要在多种转换之间选择，则会引起错误。例如：

```
void my_swap(long&,long&);             //交换两个整数的值
void my_swap (double&,double&);        //交换两个double数的值
void my_swap (string&,string&);        //交换两个string对象的值
int a=5,b=4;
my_swap(a,b);                          //错：无法确定作long型还是double型的转换
```

下面程序演示了函数重载的使用。

```
/************************************************************
    File name    : f0107.cpp
    Description  : 演示函数重载的使用
************************************************************/
```

```cpp
#include<iostream>
#include<string>
usingnamespace std;
void my_swap(int&, int&);        //函数声明
void my_swap(double&, double&);
void my_swap(string&, string&);
void my_swap(longint&, longint&);
int main()
{
    int ix = 4, iy = 5;
    double dx = 4.1, dy = 5.1;
    string strx = "string x", stry = "string y";
    char cx = 'x', cy = 'y';

    cout <<"Before swapping\n";
    cout <<"ix="<< ix <<",iy="<< iy << endl;
    cout <<"dx="<< dx <<",dy="<< dy << endl;
    cout <<"strx="<< strx <<",stry="<< stry << endl;
    cout <<"cx="<< cx <<",cy="<< cy << endl;

    my_swap(ix, iy);              //严格匹配my_swap(int&,int&)
    my_swap(dx, dy);              //严格匹配my_swap(double&, double&)
    my_swap(strx, stry);          //严格匹配my_swap(string&, string&)
    my_swap((int&)cx, (int&)cy);  //显式类型转换后匹配my_swap(int&,int&)
    cout <<"After swapping:\n";
    cout <<"ix="<< ix <<",iy="<< iy << endl;
    cout <<"dx="<< dx <<",dy="<< dy << endl;
    cout <<"strx="<< strx <<",stry="<< stry << endl;
    cout <<"cx="<< cx <<",cy="<< cy << endl;
    return 0;
}
void my_swap(int& ia, int& ib)
{
    int temp = ia;
    ia = ib;
    ib = temp;
}
void my_swap(double& da, double& db)
{
    double temp = da;
    da = db;
    db = temp;
}
void my_swap(string& stra, string& strb)
{
    string str_temp = stra;
    stra = strb;
    strb = str_temp;
```

```
}
void my_swap(longint& la, longint& lb)
{
    longint temp = la;
    la = lb;
    lb = temp;
}
//=====================================================================
```

程序的输出结果为：

```
Before swapping:
ix=4, iy=5
dx=4.1, dy=5.1
strx=string x, stry=string y
cx=x, cy=y
After swapping:
ix=5, iy=4
dx=5.1, dy=4.1
strx=string y, stry=string x
cx=y, cy=x
```

程序定义了 4 个重载函数，第 19～21 行都能严格匹配一个重载函数，满足函数参数匹配规则。第 22 行的实参类型是 char，在 C++中 char 可以转换为 int 型，也可以转换为 long int 型，如果没有显式类型转换，则无法确定是哪种转换而导致编译错误。

4. 带默认参数值的函数

C++允许在函数声明中为参数给定默认值，在调用函数时，编译器按照从左向右的顺序将实参与形参匹配，若未指定足够的实参，则编译器对未匹配的形参使用函数声明中给定的默认值。

函数参数设置默认值必须遵循以下规则：

(1) 默认参数值只能在函数声明中设定，只有当程序中没有函数声明，默认值才可以在函数定义的头部设定。

(2) 设定函数的默认参数值时必须按照从右向左的顺序设定，即设定了默认值的参数，其右边的参数都必须设定默认值。

下面程序演示了带默认参数值的函数的使用。

```
/***************************************************************
    File name    : f0108.cpp
    Description  : 演示带默认参数值的函数的使用
***************************************************************/
#include<iostream>
#include<iomanip>
usingnamespace std;

void print(int* parry, int count, int num = 5);//函数声明
int main()
{
```

```cpp
    int i_arry[15], count;
    cout <<"Please input some number:\n";
    count = 0;
    while (cin >> i_arry[count])        //输入失败则退出输入
    {
        count++;
        if (count >= 15)                //当输入数据超出数组容量，则退出输入
            break;
    }
    cout <<"Use default parameter:\n";
    print(i_arry, count);
    cout <<"Don't use default parameter:\n";
    print(i_arry, count, 6);
    return 0;
}
void print(int *parry, int count, int num)
{
    for (int i = 0; i < count; i++)
    {
        cout << setw(5) << *(parry + i);
        if ((i + 1) % num == 0)
            cout <<"\n";
    }
    cout <<"\n";
}
//================================================================
```

程序的输出结果为：

```
Please input some number:
1 2 3 4 5 6 7 8 9 a
Use default parameter:
    1    2    3    4    5
    6    7    8    9
Don't use default parameter:
    1    2    3    4    5    6
    7    8    9
```

程序中第 10～15 行是数据输入程序代码，如果输入失败则退出输入循环。当输入的数据与接收数据的变量类型不一致时，就会导致输入失败。在输入数据时，最后输入了一个字符 a，而非整数。这会使得 cin 输入操作返回一个无效状态，程序可以利用 cin 的状态来检测是否需要继续输入。

程序中定义了带默认参数值的 print 输出函数，其默认参数值控制输出时每行输出的整数个数。如果使用默认值，则每行输出 5 个整数，否则由调用函数时的实参控制每行输出的整数个数。

默认参数值和函数重载都是为了简化编程，函数重载也可以实现默认参数值的功能。当解决一个问题时，选择使用默认参数值还是函数重载主要依据以下规则：

(1) 如果两个重载函数完成基本相同的工作，只不过参数个数不同而已，则应该选择使用

默认参数值。前面程序中的 print 函数，也可以使用两个重载函数来实现，如：

```cpp
void print(int *parry, int count);              //每行输出 5 个整数的函数
void print(int *parry, int count, int num);     //每行输出 num 个整数的函数
```

显然用两个重载函数实现 print 函数不如用一个带默认参数值的 print 函数简单、合理。

(2) 如果需要一个参数的值来确定不同的操作，则应该用函数重载比较好。例如：

```cpp
/******************************************************************
    File name    : f0109.cpp
    Description  : 演示默认参数值和重载函数的使用
******************************************************************/
#include<iostream>
#include<iomanip>
usingnamespace std;
int sum_arry(int arry[], int count)
{
    int sum = 0;
    for (int i = 0; i < count; i++)
        sum += arry[i];
    return sum;
}
void print(int arry[], int count, bool flag = true)
{
    if (flag)
    {
        for (int i = 0; i < count; i++)
            cout << setw(5) << arry[i];
    }
    else
    {
        cout << setw(10) << sum_arry(arry, count);
    }
}
int main()
{
    int arry[] = { 1, 2, 3, 4, 5, 6, 7, 8, 9 };
    int count = 9;
    if (sum_arry(arry, count) > 100)
        print(arry, count);
    else
        print(arry, count, false);
    return 0;
}
//=================================================================
```

第 11 行定义了 print 函数，因为程序中没有 print 函数的声明语句，所以参数 flag 的默认值在函数定义中给出。

该程序根据数组所有元素之和来决定是输出数组的所有元素还是输出数组所有元素之

和。显然 print(arry,count) 和 print(arry,count,false) 两种函数调用都是输出工作，但内容却完全不同。对于这种需要通过一个参数来确定完成什么工作的情形，使用重载函数实现更合理，也更符合程序的逻辑性和可维护性。例如：

```cpp
/*******************************************************************
    File name   :   f0110.cpp
    Description :   演示默认参数值和重载函数的使用
*******************************************************************/
#include<iostream>
#include<iomanip>
usingnamespace std;
int sum_arry(int arry[], int count)
{
    int sum = 0;
    for (int i = 0; i < count; i++)
        sum += arry[i];
    return sum;
}
void print(int arry[], int count)
{
    for (int i = 0; i < count; i++)
        cout << setw(5) << arry[i];
}
void print(int value)
{
    cout << setw(10) << value;
}
int main()
{
    int arry[] = { 1, 2, 3, 4, 5, 6, 7, 8, 9 };
    int count = 9;
    int value = sum_arry(arry, count);
    if (value > 100)
        print(arry, count);
    else
        print(value);
    return 0;
}
//================================================================
```

1.3 输入/输出流的基本知识

如果没有数据，计算就没有任何意义，任何程序设计语言都应该提供数据输入和结果输出展示的能力。C++本身没有定义输入输出操作，而是通过 C++标准库中的输入/输出流（I/O stream）类库完成数据的输入输出。本节将简单介绍标准输入/输出流的基本知识、文件流的基本知识。

1.3.1 标准 I/O 流

输入操作可以理解为从输入流对象中提取数据,称为提取操作。C++使用标准输入流对象 cin 来实现从键盘读取数据,其一般形式如下:

    cin>>变量 1>>变量 2>>…>>变量 n;

其中,cin 是预定义的标准输入流对象,读作"see-in",是 character input 的缩写。">>"是输入操作符,也称为提取运算符。C++通过提取运算符从 cin 输入流中获取各种类型数据,赋值给相应类型的变量。输入流的数据输入**默认分隔符**包含空格、'\n'、Tab 键。

输出操作可以理解为将数据插入到输出流对象中,称为插入操作。C++使用预定义的输出流对象 cout 来实现数据输出到屏幕上,其一般形式如下:

    cout<<表达式 1<<表达式 2<<…<<表达式 n;

其中,cout 是预定义的标准输出流对象,读作"see-out",是 character output stream 的缩写。"<<"是输出操作符,也称为插入运算符。C++通过插入运算符将各种数据插入到输出流对象,由输出流对象根据数据类型按照正确的方式输出到屏幕上。输出换行可以通过 endl 控制符和转义字符 '\n' 实现。控制符 endl 不仅仅有换行功能,同时会刷新输出缓冲区;转义字符 '\n' 只有单纯的换行功能。

标准输入输出流对象在 iostream 文件中预定义,只要在程序文件开始包含此文件就可以使用标准输入流对象 cin 和标准输出流对象 cout。

1.3.2 文件流

数据一般都是保存在文件中,因此,文件操作是程序设计中非常重要的部分。C++通过标准库提供文件操作。程序设计时,只要以流的概念来处理文件即可。

文件可以分为文本文件和二进制文件。文本文件中任何内容总是与字符码表(如 ASCII 码)对应,每个字符都由一个相应的字符码表示。二进制文件不硬性规定与字符码表的对应关系,将内容看成是 0/1 二进制串,每个数据内容都以其内存中的二进制值表示。操作二进制文件时,虽然是以字节为单位,但不考虑字符的识别,更不需要数据类型的识别。

C++中通过文件流对象来实现文件操作,程序中可以使用输入文件流 ifstream 对象从文件中读取数据,使用输出文件流 ofstream 对象输出数据到文件中。ifstream 类和 ofstream 类在 fstream 文件中定义,使用文件流对象操作文件时必须包含此文件。

文件操作流程可以分成以下几步:
(1) 打开文件,并判定文件是否打开成功;
(2) 读取数据/写数据到文件;
(3) 关闭文件。

1. 打开文件

打开文件的方式就像定义一个对象,格式如下:

```
ifstream fin(filename, openmode=ios::in);
ofstream fout(filename, openmode=ios::out);
```

其中,**filename** 表示将要打开的文件名,可以包含完整的路径名。openmode 表示打开文件的方式。文件流打开文件的方式有多种,具体可以查阅第 7 章有关文件的介绍。其中输入文件流的默认方式为读取数据方式(ios::in),输出文件流的默认方式为输出数据方式(ios::out)。

打开文件后,一定要检查文件是否打开成功才能进行文件的读写操作,否则当文件打开不成功就进行文件操作,会导致文件操作失败。判定文件是否打开成功的语句如下:

```
//打开读取数据的文件                        //打开输出数据的文件
ifstream fin("D:\\data.txt");          ofstream fout("D:\\out.txt");
if (!fin)                              if (!fout)
{                                      {
    cout<<"Can't Open file!\n";            cout<<"Can't Open file!\n";
    exit(0);                               exit(0);
}                                      }
```

上面语句中,通过定义文件流对象直接打开文件。如果文件流对象成功打开文件,则文件流对象值为非 0,其与文件建立了关联关系。如果文件流对象打开文件不成功,则文件流对象值为 0。可以直接使用"!fin"和"!fout"来判断文件流对象打开文件是否成功。也可以利用文件流的 open 函数打开文件,将文件流对象与具体文件关联,具体方式可查阅第 7 章相关知识。

2. 读取数据

文本文件和二进制文件的读取方式不同。本章只介绍文本文件的读取方法,二进制文件的数据读取方法请参考第 7 章相关内容。

C++中有两种方式读取文本文件数据:一种是使用提取运算符直接从输入文件流中读取数据;另一种是使用全局函数 getline 每次读取文件的一行数据,然后使用字符串流对象分离数据。下面分别用程序演示两种文本文件数据读取方法。

读取文件中单词的要求:当文件中不超过 100 个单词时,则读取文件中的所有单词;当文件中超过 100 个单词时,只读取文件的前面 100 个单词。假设 D 盘根目录有文本文件 data.txt,文件中存放的数据如下:

```
This is a C++ program
for file's operation.
```

使用提取运算符从输入文件流中读取单词的程序如下:

```
/*****************************************************************
    File name    : f0111.cpp
    Description  : 提取运算符读取文本数据
*****************************************************************/
#include<iostream>
#include<fstream>
#include<string>
usingnamespace std;
int main()
{
    string str_arry[100];
    int count;
    //定义输入文件流对象,并打开文件
    ifstream in("D:\\data.txt");
```

```cpp
        if (!in)
        {
            cout <<"Can't open the file!"<< endl;
            exit(0);
        }
        //读取数据
        count = 0;
        while (in >> str_arry[count])
        {
            count++;
            if (count > 100)    //当读取的单词个数达到100个，则不再读取单词
                break;
        }
        for (int i = 0; i < count; i++)
        {
            cout << str_arry[i] <<" ";
        }
        cout << endl;
        return 0;
    }
    //=====================================================
```

程序的输出结果为：

```
This is a C++ program for file's operation.
```

程序中 in 是 ifstream 流对象，并且已经打开了一个包含字符串数据的文件 data.txt。第 18 行使用提取运算符将输入文件流对象 in 的数据读取到 string 数组 str_arry 的第 count 个元素中。如果读数据到文件尾，则会导致读取数据失败，in 提取操作失败，从而退出循环。当读取的单词个数达到了数组的最大存放档次个数，则退出读取数据循环。上述代码段完成了从包含字符串数据的文本文件中读取不超过 100 个单词的功能。

使用全局函数 getline 读取文件的程序如下：

```cpp
/******************************************************************
    File name    : f0112.cpp
    Description  : 全局函数 getline 和字符串流读取文本数据
******************************************************************/
#include<iostream>
#include<fstream>
#include<string>
#include<sstream>
usingnamespace std;
int main()
{
    string str_arry[100];
    string str;
    int count;
    //定义输入文件流对象，并打开文件
    ifstream in("f0111_data.txt");
```

```cpp
    if (!in)
    {
        cout <<"Can't open the file!"<< endl;
        exit(0);
    }
    //读取数据
    count = 0;
    while (getline(in, str))
    {
        istringstream sin(str);
        while (sin >> str_arry[count])
        {
            count++;
            if (count > 100)
                break;
        }
        if (count > 100)
            break;
    }
    for (int i = 0; i < count; i++)
    {
        cout << str_arry[i] <<" ";
    }
    cout << endl;
    return 0;
}
//================================================
```

程序的输出结果为：

```
This is a C++ program for file's operation.
```

程序中 in 为 ifstream 流对象，str 是 string 类对象，存放从文件中读取的一行数据。getline 读取失败时返回错误状态，从而退出循环。getline 将数据按照字符串的方式读取，每次读取一行数据。sin 是输入字符串流对象，用于从 string 类对象中分离数据，使用字符串流对象时必须包含 sstream 头文件。有关字符串流的详细介绍可以参阅第 8 章的相关内容。

3. 输出数据

文本数据可以使用插入运算符将数据插入到输出文件流来完成数据到文件的输出，文本文件的输出还可以通过流状态或流的成员函数来进行格式化输出。例如：

```cpp
/***************************************************************
    File name    : f0113.cpp
    Description  : 文本数据输出到文件
***************************************************************/
#include<iostream>
#include<fstream>
usingnamespace std;
int main()
```

```
    {
        ofstream fout("data.txt");
        if (!fout)
        {
            cout <<"Can't Open file!\n";
            exit(0);
        }
        int i_arry[10] = { 1, 2, 3, 4, 5, 6, 7, 8, 9, 11 };
        for (int i = 0; i < 10; i++)
        {
            fout << setw(5) << i_arry[i];
        }
        return 0;
    }
//================================================================
```

上面程序中，fout 是 ofstream 流对象，并且已经打开一个用于保存数据的文本文件。setw 是一种流状态控制函数，在此控制每个整数占 5 列输出。

4. 关闭文件

关闭文件的方法有两种：一种是不做任何操作，由文件流对象的析构函数自动关闭文件。当程序离开文件流对象的作用域时，对象会被销毁，它所关联的文件也会被关闭。使用析构函数关闭文件可以最大限度地降低两类错误出现的概率：在打开文件之前或关闭文件之后使用文件流对象。另一种是调用文件流对象的 close 成员函数。例如：

```
    ifstream  fin;
    //...
    fin >> num;                         //错：没有用 fin 打开文件
    //...
    fin.open(name, ios_base::in);       //打开文件
    //...
    fin.close();
    //...
    fin >> value;                       //错：文件已经关闭
```

1.3.3 流状态

C++的 I/O 流类库主管数据类型的识别和沟通操作系统，全面负责把流中的数据送到对应的设备上。I/O 流类库提供了一种使用格式控制符进行格式化输出的方法，如对齐、宽度限定、精度设置、数制等显示形式。

1. 常用的流状态（表 1-3）

表 1-3 常用流状态

流状态符	含　　义	特　　性
showpos	为正数之前（包括 0）显示+号	持久，noshowpos 取消
showbase	为八进制数加前缀 0，为十六进制数加前缀 0x	持久，noshowbase 取消
uppercase	十六进制格式字母用大写显示（默认为小写）	持久，nouppercase 取消

流状态符	含 义	特 性
showpoint	浮点输出，即使小数点后都为0也显示小数	持久，noshowpoint 取消
boolalpha	逻辑值1和0用 true 和 false 显示	持久，noboolalpha 取消
left	左对齐(填充字符在右边)	持久，right 取消
right	右对齐(填充字符在左边)	持久，left 取消
dec	十进制显示整数	持久，hex 或 oct 取消
hex	十六进制显示整数	持久，dec 或 oct 取消
oct	八进制显示整数	持久，dec 或 hex 取消
fixed	使用定点显示，精度为小数点之后数字个数	持久，scientific 取消
scientific	使用尾数和指数表示方式，尾数总在[1:10]之间，也就是小数点之前只有一个非0数字，精度为小数点之后的数字个数	持久，fixed 取消

例如：

```
cout << showpos << 12 << " " << 34 << noshowpos << endl;
                                        //输出:+12  +34
cout << hex << 18 <<" "<< showbase << 16 << endl;
                                        //输出:12  0x10
cout << dec << left << setw(5) << 5 << setw(5) << 6 << endl;
                                        //输出:5    6
cout << right << setw(5) << 5 << setw(5) << 6 << endl;
                                        //输出:    5    6
cout << showpoint << 12.0 <<" "<< 13.0 << endl;
                                        //输出:12.0000  13.0000
cout << 14.01 <<" "<< 15.04 << endl;
                                        //输出:14.0100  15.0400
cout << fixed << 14.01 <<" "<< 15.04 << endl;
                                        //输出:14.010000 15.040000
cout << scientific << 14.01 <<" "<< 15.04 << endl;
                                        //输出:1.401000e+001 1.504000e+001
```

2. 有参数的三个常用流状态

有三个常用的流状态是带参数的，如表1-4所示。

表1-4 带参数的流状态

流状态	含 义	特 性
width(int)	设置显示宽度	一次性操作
fill(char)	设置填充字符	持久，fill(' ') 取消
precision(int)	设置有效位数(普通显示方式)或精度(fixed 或 scientific 方式)	持久

这三个有参数的流状态是流类的成员函数，需要由 cout 流对象调用才能使用。例如：

```
cout.width(6);
cout << 5 << endl;              //输出:□□□□□5
cout.fill('x');
cout.width(3);
cout << 6;
cout.width(4);
cout << 7 << endl;              //输出:xx6xxx7
```

```
cout.width(4);
cout.fill(' ');
cout << 7 << endl;                        //输出:□□□7
cout.precision(4);
cout << 12.011234 << endl                 //输出:12.01
cout << fixed << 13.0212345 << endl;      //输出:13.0212
```

第 2 和 7 行中，□表示空格。在默认显示格式下，实数的精度是指其有效数字个数，在 fixed 显示格式下，实数的精度是指小数点之后的数字个数。Precision 设置精度是持久的，所以上面代码中第 12 行显示为 12.01，只有两位小数，第 13 行显示 4 位小数。

3. 与插入运算符(<<)连用的设置方式

C++中提供了一种<<连用的方式(表 1-5)，在使用此方式时，需要包含头文件 iomanip。

表 1-5 与插入运算符连用的流状态

流状态	含 义	特 性
setw(int)	设置显示宽度	一次性操作
setfill(char)	设置填充字符	持久，setfill(' ')取消
setprecision(int)	设置有效位数(普通显示方式)或精度(fixed 或 scientific 方式)	持久

例如：

```
cout <<setw(5)<< 5 << endl;                  //输出:□□□□5
cout<<setfill('x')<<setw(3)<< 6;
cout <<setw(4)<< 7<<endl;                    //输出:xx6xxx7
cout << setfill(' ') << setw(4) << 7 << endl; //输出:   7
cout <<setprecision(4)<< 12.011234 << endl;  //输出:12.01
cout << fixed << 13.0212345 << endl;         //输出:13.0212
```

1.3.4 文件操作实例

【例题】 现有文本文件 data.txt 保存了不超过 100 个整型数。要求读取文件中的所有整数，并输出到屏幕上，要求每行输 8 个整型数，同时，以相同的格式输出到文本文件 result.txt 中。

```
/*****************************************************************
    File name    :  f0114.cpp
    Description  :  演示文件以及输入输出的基本操作
*****************************************************************/
#include<iostream>
#include<fstream>
#include<iomanip>
usingnamespace std;
int main()
{
    ifstream fin("data.txt", ios_base::in);   //打开输入文件
    //判断文件是否打开成功
    if (!fin)
    {
        cout <<"Can't open file: data.txt"<< endl;
        exit(0);
```

```
    }
    ofstream fout("result.txt", ios_base::out);  //打开输出文件
    //判断文件是否打开成功
    if (!fout)
    {
        cout <<"Can't open file: result.txt"<< endl;
        exit(0);
    }
    //读取数据
    int i_arry[100];
    int count = 0;
    while (fin >> i_arry[count])
    {
        count++;
        if (count >= 100)
            break;
    }
    //输出到屏幕和文件中
    for (int i = 0; i < count; i++)
    {
        cout << setw(5) << i_arry[i];         //输出到屏幕
        if ((i + 1) % 8 == 0)
            cout <<"\n";

        fout << setw(5) << i_arry[i];         //输出到文件
        if ((i + 1) % 8 == 0)
            fout <<"\n";
    }
    return 0;
}
//==================================================================
```

程序首先打开文件并对打开文件操作进行检测。当文件打开成功，然后读取文件数据，根据输入文件流状态来判断是否需要继续读文件。最后输出时可以看出，标准输出与文件输出的操作形式完全一样，只是标准输出使用标准输出流对象 cout，文件输出使用输出文件流对象。

1.4 异常基本知识

编程就是预先设定一个动作序列让计算机执行。程序产生什么结果应该在预料之中，即使出现错误，也应该有一个预定的处理方案。在结构化框架中，一般对错误的处理方法有如下几种。

(1)遇到错误，立即终止程序运行。这是最粗暴的一种错误处理方法。例如：在本书样例程序中，对打开文件失败的错误就是简单地终止程序运行。

(2)返回一个表示错误的值给上层函数。这种方法是将函数中的错误交给函数使用者来处理。这样做的结果是导致每次调用函数都必须进行错误检查，增加程序设计工作量，程序规模也会不知不觉中庞大起来。例如，判断日期是否合法时，只把判断结果以某些值中的一个返回给上层函数，每个值表示一种错误信息。

(3) 返回一个合法值，然后通过全局变量设置错误状态。这种方式实际上与第 2 种类似，只是将错误信息集中到了一个专门用于表示错误状态的全局变量中。

(4) 调用事先准备好的错误处理函数，让其去决定采取什么样的操作。这种错误处理函数处理完后便要返回到调用处。但很多错误都不是现场能恢复的，因此，就地做了错误处理也不一定能让程序继续运行，最后也不得不粗暴地停止程序。

C++使用了一种新的错误处理机制——异常处理（exception handing）。异常处理机制是用于管理程序运行期间出现非正常情况的一种结构化方法，是当面向对象理念开始进入编程实战时就产生的，专门针对抽象编程中一系列错误处理的方法。C++的异常处理将异常的检测和异常处理分离，增加了程序的可读性。异常处理也是提高程序健壮性的重要手段。

C++异常处理技术的使用方法可以描述为以下三步：

① 框定异常（try 语句）

在程序中，框定认为可能发生错误的语句序列，这些序列是异常发生的根源，如不框定异常，则捕获不到任何异常。

② 定义异常处理（catch 语句）

将出现异常后的处理过程放在 catch 语句块中，当抛出异常时，就能因类型匹配而捕获异常。

③ 抛弃异常（throw 语句）

在可能产生异常的语句中进行错误检测，如果有错误，就抛出异常。

框定异常和处理异常时在同一个函数中完成，抛出异常既可以和框定异常、处理异常在同一个函数，也可以跨越函数。下面例子展示了异常处理的基本使用。

```cpp
/******************************************************************
    File name:      f0115.cpp
    Description:    演示异常的基本操作
******************************************************************/
#include<iostream>
#include<string>
using namespace std;
int strToint(const string& str)
{
    int num = 0;
    for (unsignedint i = 0; i < str.length(); i++)
    {
        if ((str.at(i) >= '0') && (str.at(i) <= '9'))
        {
            num = num * 10 + str.at(i) - '0';      //将数字字符转换成数字
        }
        else
        {
            throw str.at(i);
        }
    }
    return num;
}
int main()
{
```

```cpp
        string str;
        int i_value;
        cout <<"Please input a number string:\n";
        cin >> str;
        try
        {
            i_value = strToint(str);
            if (i_value > 10000) throw i_value;
        }
        catch (char)
        {
            cout <<"Input string includes illegal character: "<<char<< endl;
            i_value = -1;
        }
        catch (int)
        {
            cout <<"number is too big: "<< i_value << endl;
        }
        cout << i_value << endl;
        return 0;
    }
    //=========================================================
```

程序第一次运行的输出结果为：

```
Please input a number string:
12
12
```

程序第二次运行的输出结果为：

```
Please input a number string:
1234a5
Input string includes illegal character: a
-1
```

程序第三次运行的输出结果为：

```
Please input a number string:
12345
number is too big: 12345
12345
```

程序中异常抛出位置既可以在 try 语句块，又可以在 try 语句调用的函数中。字符串转换为整型数的函数参数为 string 对象的常引用，其目的是限定函数对参数只有读的访问权限，不具有写的权限，这种使用方法在 C++中非常常见。

练　习

1. 找出包含 20 元素的整型数组中的最大值、最小值，并计算数组所有元素的平均值，按要求完成如下工作：

(1)使用指针作为函数参数设计一个函数实现以上功能。

(2)使用引用作为函数参数设计一个函数实现以上功能。

要求在 main 函数中使用随机数初始化数组,并在 main 函数中输出数组的最大值、最小值和平均值。

注意:思考指针和引用作为函数在函数实现和函数功能方面的区别与共同点。

2. 判断字符串 s1 中是否包含字符串 s2,如果包含则返回第 1 个 s2 在 s1 中的第 1 个字符位置,如果 s1 中包含多个 s2 子串,则只需找到第一个子串。按要求完成如下工作:

(1)使用 C-字符串类型表示字符串 s1 和 s2,设计一个函数实现以上功能。

(2)使用 string 类型表示字符串 s1 和 s2,并且不使用 C++标准库函数,设计一个函数实现以上功能。

(3)使用 string 类型表示字符串 s1 和 s2,使用 C++标准库函数,设计一个函数实现异常功能。

要求在 main 函数中输入两个字符串,并在 main 函数中输出 s2 在 s1 中的起始位置。

注意:思考 C-字符串与 string 类型在函数实现和函数功能方面的区别与共同点,思考不使用 C++标准库函数和使用 C++标准库函数在函数实现上的区别。

3. 使用 string 类型输入一个英文语句,试编写一个函数将语句中的所有单词转换成大写字母,并在 main 函数中输出转换后的英文语句。

4. 使用 string 类型输入一个英文语句,试编写一个函数提出语句中的所有单词,并在 main 函数中输出所有单词,要求每行输出一个单词。

5. 使用 string 类型输入若干单词,试完成如下功能:

(1)按照字典顺序对所有单词从小到大排序,并在 main 函数中输出排序后的所有单词,要求一个单词一行。

(2)根据单词长度对所有单词从小到大排序,并在 main 函数中输出排序后的所有单词,要求一个单词一行。

6. 创建三个分别包含若干个整数、若干个实数和若干个字符串的文本文件,试编写函数完成如下功能:

(1)编写函数,实现从三种不同类型数据文件中读取数据。

(2)编写函数,分别找出三种不同类型数据集合中的最大者、最小者。

(3)编写函数,将三种不同类型的数据输出到屏幕上。

注意:根据题目要求完成任务,请考虑函数重载方法、引用和指针作为函数参数。

7. 创建一个包含不超过 100 个整型数的文本文件 data.txt,试完成如下功能:

(1)编写一个函数,从文件中读取所有整型数。

(2)在 main 函数中输出所有整型数。格式要求每行输出 8 个整数,每个整数占 5 列。

(3)编写一个函数,删除所有整型数中包含数字 2 和 4 的整数。在 main 函数中输出所有剩余的整数,输出格式与(2)相同。

(4)编写一个函数,将(3)中的所有整数输出到文本文件 result.txt 中,格式与输出到屏幕的格式相同。

第 2 章 类和对象的创建

面向对象程序设计是一种不同于结构化程序设计的思想。C++通过建立类，把事物的属性（以变量的形式）以及处理事物的方法（以函数的形式）封装到一起。面向对象的程序设计思想更符合人类的认知，为大型软件的设计开发提供了一种更为有效的手段。

- **知识要点**
  理解面向对象技术的特点；
  掌握类的定义及其简单使用方法。
- **探索思考**
  面向对象方法与结构化方法的区别。
- **预备知识**
  复习结构体类型的设计和使用；
  函数参数传递方式。

2.1 从结构体到类

在 C 语言中我们已经学习过两种主要的用户自定义数据类型：枚举型（enum）和结构体（struct），此外还有联合体（union）等。它们语法的共同特征是定义的右括号(})之后要加分号(;)。可以利用用户自定义数据类型定义变量。

C 语言的结构体可以把相关的数据元素组成一个统一体，便于后续的处理。例如，下面的代码定义了一个日期结构体 Date：

```
/******************************************************************
    File name    : f0201.cpp
    Description  : 结构体的使用
******************************************************************/
#include <iostream>
using namespace std;

struct Date
{
    int year;
    int month;
    int day;
};

void main()
{
    Date today,yesterday,tomorrow;
    today.year=2015;
```

```
        today.month=7;
        today.day=17;
        yesterday.year=tomorrow.year=today.year;
        yesterday.month=tomorrow.month=today.month;
        yesterday.day=today.day-1;
        tomorrow.day=today.day+1;
        cout<<yesterday.year<<"-"<<yesterday.month<<"-"<<yesterday.day<<endl;
        cout<<today.year<<"-"<<today.month<<"-"<<today.day<<endl;
        cout<<tomorrow.year<<"-"<<tomorrow.month<<"-"<<tomorrow.day<<endl;
    }
    //==================================
```

程序的输出结果为:

```
    2015-7-16
    2015-7-17
    2015-7-18
```

结构体中定义了三个整型成员 year(年)、month(月)和 day(日)。

主函数中定义了 3 个结构体变量 today、yesterday 和 tomorrow。today 的年月日分别赋值为 2015、7 和 17，yesterday.day 和 tomorrow.day 通过 today.day-1 和 today.day+1 赋值。相应的日期通过 cout 语句输出到显示器上。可以看出，结构体中的成员可以在结构体的外面随意使用，这在为程序设计提供便利的同时，也极易出错。如将主程序修改为:

```
    void main()
    {
        Date today,yesterday,tomorrow;
        today.year=2015;
        yesterday.year=tomorrow.year=today.year;
        today.month=2;
        yesterday.month=tomorrow.month=today.month;
        today.day=28;
        yesterday.day=today.day-1;
        tomorrow.day=today.day+1;
        cout<<yesterday.year<<"-"<<yesterday.month<<"-"<<yesterday.day<<endl;
        cout<<today.year<<"-"<<today.month<<"-"<<yesterday.day<<endl;
        cout<<tomorrow.year<<"-"<<tomorrow.month<<"-"<<tomorrow.day<<endl;
    }
    //==================================
```

程序的输出结果为:

```
    2015-2-27
    2015-2-27
    2015-2-29
```

这里，忽略了 2 月 28 号的明天不是 2 月 29 日，而是 3 月 1 日。而且，本应打印 today 的信息，程序却错误地将 today.day 写成 yesterday.day。用户当然可以通过细心编程，避免发生错误；也可以通过认真检查，发现并纠正这些问题。但是，如果程序设计能够提供一种方法，可以保证不发生类似的错误，岂不更好！

上述的错误可以通过建立函数加以避免。通过传递函数参数，对 Date 对象的值进行处理。在 C++中，函数参数的传递方法有三种：

(1) 值传递(pass by value，C 语言已学过)。
(2) 指针传递(pass by pointer，C 语言已学过)。
(3) 引用传递(pass by reference，C++新增)。

下面的程序展示了几种函数参数传递的情况：

```cpp
/******************************************************************
    File name    :  f0202.cpp
    Description  :  函数参数的传递方式
******************************************************************/
#include <iostream>
using namespace std;

struct Date
{
    int year;
    int month;
    int day;
};
void init_date1(Date date,int year,int month,int day)
{
    date.year=year;
    date.month=month;
    date.day=day;
}
void init_date2(Date* pdate,int year,int month,int day)
{
    pdate->year=year;
    pdate->month=month;
    pdate->day=day;
}
void init_date3(Date& date,int year,int month,int day)
{
    date.year=year;
    date.month=month;
    date.day=day;
}
void print(Date& date)
{
    cout<<date.year<<"-"<<date.month<<"-"<<date.day<<endl;
}
void main()
{
    Date today,yesterday,tomorrow;
    init_date1(today,2015,2,28);
    print(today);
```

```
        init_date2(&yesterday,2015,2,27);
        print(yesterday);
        init_date3(tomorrow,2015,3,1);
        print(tomorrow);
}
//=====================================
```

程序的输出结果为:

```
-858993460--858993460--858993460
2015-2-27
2015-3-1
```

init_date1 函数是值传递,可以看出,利用值传递只是改变了形参的内容,无法完成对实参值 today 的改变。init_date2 函数是指针传递,通过指针可以改变 yesterday 的内容。指针的参数传递仍然是值传递,只不过值传递的内容是 yesterday 的指针。init_date3 函数是引用传递,形参就是实参,对形参的修改就是对实参的修改。三种参数传递方式中,引用的方式更适合于对实参内容的修改。打印函数 print 是以对象为单位进行处理,避免了把 today.day 错写成 yesterday.day 的可能。

可以看到,这些函数本质上与结构 Date 没有直接关系,是通过参数的传递来改变实参的值。实际上,如果把 Date 作为一个整体,我们更希望这些函数与 Date 类直接相关,也就是说,这些函数可以直接对 Date 中的 year、month 和 day 进行处理,而不是通过参数传递的方式。

通常意义下,C 语言提供的结构体(struct)没有提供这样的功能。C++提供了类(class)这样一种用户自定义数据类型的方式,可以将变量和函数统一进行处理。

2.2 类声明和对象定义

C++的类既可以包括数据成员,也可以包含对数据成员进行各种操作的成员函数。下面的例子是将 Date 定义成类的程序。

```
/****************************************************************
    File name   : f0203.cpp
    Description : 类的定义
****************************************************************/
#include <iostream>
using namespace std;

class   Date
{
private:
    int year;
    int month;
    int day;
public:
    inline void init_date(int y,int m,int d);
    bool isLeapYear()
    {
        if( year%400==0 || (year%4==0 && year%100!=0))
```

```cpp
            return true;
        else
            return false;
    }
    void print()
    {
        cout<<year<<"-"<<month<<"-"<<day<<endl;
    }
};
void Date::init_date(int y,int m,int d)
{
    year=y;
    month=m;
    day=d;
}

void main()
{
    Date today,yesterday,tomorrow;
    today.init_date(2015,2,28);
    yesterday.init_date(2015,2,27);
    if(today.isLeapYear())
        tomorrow.init_date(2015,2,29);
    else
        tomorrow.init_date(2015,3,1);
    yesterday.print();
    today.print();
    tomorrow.print();
}
//==================================
```

程序的输出结果为：

```
2015-2-27
2015-2-28
2015-3-1
```

类与结构体相似，也是程序员定义新数据类型的方法，其右括号后面也要加一个分号(;)。既然类是一种新的数据类型，因此也可以通过类定义变量，这里的变量通常称为对象。

2.2.1 类的数据成员和成员函数

上例中，关键字 class 表示类，Date 是类名，一般首字母用大写字母表示，以示与对象名的区别。声明的日期类 Date，包含 3 个数据项：year、month 和 day，分别表示年、月和日，均被定义成整型量 int。这些在类中定义的数据项称为数据成员，也称为成员变量或属性。

类 Date 中也包含了 3 个函数：

```cpp
void init_date(int y,int m,int d);
bool isLeapYear();
void print();
```

这些在类中定义的函数称为成员函数，也称为方法。init_date(int y,int m,int d)用于为 3 个数据成员赋值，isLeapYear()用于判断当前对象是否为闰年，如果当前的年份是闰年，返回 true，否则返回 false。print()则用于输出日期到显示器。

可以看到，在成员函数中可以直接访问数据成员，而不需要利用参数传递。这些数据成员是利用类定义的对象，通过 public 的成员函数来处理对象的属性。例如，通过 today.init_date (2015,2,28)，将对象 today 的数据成员 year、month 和 day 赋值为 2015、2、28。

主程序也做了相应修改。当 if(today.isLeapYear()) 为 true 时，tomorrow 被初始化成 (2015,2,29)，否则初始化成(2015,3,1)。这里需要注意，不能用 if(tomorrow.isLeapYear())进行判断，因为此前 tomorrow 的内容尚未意义。

成员函数可以定义在类声明的内部，如 void print()，也可以先在类声明内部声明一个函数 void init_date(int y,int m,int d)，然后在类的外面定义该函数。此时，必须用作用域运算符::来通知编译系统该函数所属的类。定义的格式如下：

```
<函数类型><类名>::<函数名>(<形式参数表>)
{
    //函数体
}
```

只要在类定义中包含的成员函数，就默认声明为内联(inline)函数。类内声明，类外定义的成员函数如果也想成为内联函数，就需要在类内声明时，在函数返回值类型前面加上关键字 inline，在类外定义时就不需要再加上 inline 了。编译器对内联函数的处理与普通函数的处理不同。普通函数调用需要额外的运行开销，即将参数和寄存器内容压入栈，并且在函数间切换控制时也需要花费时间。C++提供了内联函数，可以避免函数调用的开销。内联函数不会被调用，而是编译器将其代码复制到每个调用点上，避免了函数调用所花的开销。编译是否真的将这些成员函数处理为内联函数，还要看函数是否足够简单，是否包含了不适于内联运行的循环指令等。内联函数对于短函数而言是适合的，但并不适合在程序中多次调用的长函数。在此情况下使用内联函数将会急剧增加可执行代码的长度。因此，C++允许编译器对过长的函数忽略 inline 关键字。

2.2.2 访问权限

在类的定义中使用了 private(私有)和 public(公有)两个关键字，它们为成员规定了访问权限。除了 private 和 public 之外，还有 protected(保护)。类中默认的访问权限是 private，如果将程序中的 private 去掉，定义的 year、month 和 day 将接受类的默认访问权限，仍为 private。访问权限一经设定，后续定义的成员都将遵循该访问权限，直到重新设定访问权限为止。

面向对象程序设计的一个关键特性在于数据隐藏，即将数据成员隐藏在类的内部，不让类外的函数随意访问而造成错误。类中将需要隐藏的数据成员或者成员函数的访问权限定义成 private(私有)或者 protected(保护)，于是类外的函数就不能访问它们了。而定义成 public(公有)的数据成员和成员函数则可以在类的外面被访问。目前，private 和 protected 的作用没有区别，当学到第 5 章类的继承时，二者的区别才会显现出来。

这里特别要强调面向对象程序设计中的两个视角：一个是类的设计者视角，一个是类的使用者视角。

类的设计者在类成员函数中可以直接使用数据成员及成员函数。类的使用者则需要首先定义类对象，然后通过具有 public 访问权限的成员函数或者数据成员对定义的类对象进行操作。但是，为了保持数据隐藏，我们建议将数据成员定义为 private 或者 protected，对数据成员的操作，则通过定义访问权限为 public 的成员函数来完成，这些成员函数称为接口函数。例如，init_date(int y,int m,int d) 函数中直接使用 year=y，其意义是对于任何定义的对象，其数据成员 year 用参数 y 进行赋值。month 和 day 也是如此。这样，数据成员的操作是通过成员函数 init_date 来完成的。对象可以直接使用 public 的成员函数，如

```
Date today;
today.init_date(2015,10,19);
```

来实现对 today 的 year、month 和 day 的赋值，而不能直接使用

```
today.year=2064;        //error
```

因为 year 是 private 访问权限，不能直接对其赋值。

在类定义之外，例如在主函数中，首先要创建一个 Date 对象 today。由于 init_date(int y,int m,int d) 函数的访问权限是 public，所以可以用 today.init_date(2015,2,28) 来对 today 进行赋值操作。

类的声明类似于设计了一张图纸，利用图纸可以制造工件。一张图纸并不是工件，只有根据图纸加工成工件，这时图纸的设计才有了具体的实现。同样，类本身不是任何对象，需要通过定义变量才能创建对象。创建对象也叫实例化，对象也称为实例变量，或者简称变量。

在主函数 main() 中定义了 3 个 Date 类的对象 today、yesterday 和 tomorrow：

```
Date today,yesterday,tomorrow;
```

定义的类对象可以访问 public 的数据成员和成员函数。f0203 程序的主函数中通过调用定义为 public 的成员函数 init_date() 和 print() 对 3 个对象赋初值和打印显示，isLeapYear() 则对 today 对象进行判断，如果是闰年，返回 true；否则，返回 false。

2.3 构造函数和析构函数

2.3.1 构造函数

如果把全部的人当做一个类，那么任何一个小孩子呱呱坠地，都可以认为是从人这个类中创建的对象。这些孩子出生之时，性别不同、身高不同、体重各异。同样，对于一个类创建的对象，我们也希望在对象创建的同时，就具有各自不同的属性，即数据成员应该具有相应的初始值。但情况并不像我们想象的那样。

```
/************************************************************
    File name   : f0204.cpp
    Description : 类的构造函数
************************************************************/
#include <iostream>
using namespace std;

class   Date
{
```

```cpp
    private:
        int year;
        int month;
        int day;
    public:
        void init_date(int y,int m,int d)
        {
            year=y;
            month=m;
            day=d;
        }
        bool isLeapYear()
        {
            if( year%400==0 || (year%4==0 && year%100!=0))
                return true;
            else
                return false;
        }
        void print()
        {
            cout<<year<<"-"<<month<<"-"<<day<<endl;
        }
};
void main()
{
    Date oneday;
    oneday.print();
    oneday.init_date(2015,8,3);
    oneday.print();
}
//====================================
```

程序的输出结果为：

```
-858993460--858993460--858993460
2015-8-3
```

该例子中第一个 oneday.print() 输出的日期是无意义的数字，只有通过函数：

```
oneday.init_date(2015,8,3);
```

对 oneday 进行数据赋值后，oneday.print() 输出的结果才是有意义的日期。

如何能在对象"出生"时就具有自己的属性而不是靠"后天"的赋值？C++提供了**构造函数**(constructor)，用于处理对象的初始化。构造函数是一种特殊的成员函数，与其他成员函数不同，不需要用户来调用它，而是在建立对象时自动执行。构造函数的名字必须与类名同名，以便编译系统能够识别它并把它作为构造函数处理。它不具有任何类型，不返回任何值，void 也不行。构造函数的功能是由用户定义的，用户根据初始化的需要来设计函数体和函数参数。

如果希望对不同的对象赋予不同的初始值，则需要使用带参数的构造函数，在调用不同对象的构造函数时，将不同的数据传给构造函数，以实现不同的初始化任务。

构造函数首部的一般格式为：

<构造函数名>(<形参列表>)

由于用户不能调用构造函数，因此无法采用常规的调用函数的方法给出实参。实参是在创建对象时给出的。创建对象的一般格式为：

<类名> <对象名>(<实参列表>);

例题 f0205.cpp 是带有构造函数的一个例子。

```
/*****************************************************************
    File name    : f0205.cpp
    Description  : 带参数的构造函数
*****************************************************************/
#include <iostream>
using namespace std;

class   Date
{
    int year;
    int month;
    int day;
public:
    Date(int y,int m,int d)
    {
        year=y;
        month=m;
        day=d;
    }
    void print()
    {
        cout<<year<<"-"<<month<<"-"<<day<<endl;
    }
};
void main()
{
    Date today(2015,2,28),yesterday(2015,2,27),tomorrow(2015,3,1);
    yesterday.print();
    today.print();
    tomorrow.print();
}
//===================================
```

程序的输出结果为：

```
2015-2-27
2015-2-28
2015-3-1
```

结果显示，3 个对象定义之后，便具有了各自的属性值。利用构造函数赋值较其他利用自定义的赋值函数进行赋值，更加符合对象创建的实际情况，并且是自动调用，优势明显。

2.3.2 默认构造函数

用户可以利用 f0204.cpp 声明的类,运行下面的主函数:

```
void main()
{
    Date days[100];
}
```

但是,当将 f0205.cpp 中的主函数换成该主函数,程序却不能运行。这是为什么呢?f0205.cpp 中的类提供了带有 3 个参数的构造函数,任何对象的创建,都需要 3 个参数。当用户构建具有许多元素的对象数组时,难以为每个元素提供相应的参数,此时就无法运行了。

那为什么 f0204.cpp 声明的类却可以运行呢?该类没有提供构造函数,这些没有定义构造函数的类定义对象时,是否调用构造函数呢?

C++规定,如果没有为类提供任何构造函数,编译器将自动为其生成一个默认的无参构造函数。一旦为类定义了构造函数,编译器将不再生成默认的构造函数。

系统为 Date 类提供的默认构造函数的功能如下面的程序:

```
Date()
{}
```

该构造函数的名字与类名相同,不带任何参数,没有返回值,函数体是空的,不做任何工作。

当用户需要建立许多对象,又难以逐一通过带参数构造函数为其初始化时,就必须提供默认构造函数。

首先将 f0204 的主函数换成下面的代码:

```
void main()
{
    Date days[3];
    for(int i=0;i<3;i++)
    {
        days[i].init_date(2015,4,i+1);
        days[i].print();
    }
    Date today;
    today.init_date(2015,2,28);
    today.print();
}
//====================================
```

显示的结果为:

```
2015-4-1
2015-4-2
2015-4-3
2015-2-28
```

然后,用上面的主函数替换 f0205.cpp 的主函数,并将默认构造函数:

```
    Date()
    {
        cout<<"进入构造函数"<<endl;
    }
```

加入 f0205.cpp 中 public 访问权限的位置,程序的输出结果为:

```
进入构造函数
进入构造函数
进入构造函数
2015-4-1
2015-4-2
2015-4-3
进入构造函数
2015-2-28
```

可见,建立对象数组时,调用了3次构造函数,创建 today 调用了1次构造函数,其余的结果跟上面的结果相同。

2.3.3 构造函数的重载

关于构造函数,可能有多重需求。用户需要有能够成批量初始化对象的构造函数,也需要为不同对象进行不同初始化的构造函数。这时就需要提供多个构造函数,既可以对有具体属性的对象初始化,又能对类似 days[100]这样群体对象进行统一构造,然后再分别赋值。因此,既要有默认构造函数,同时也要有带参数的构造函数,供用户选用。这些构造函数具有相同的名字,但参数的个数或参数的类型不相同。这种现象称为构造函数的重载。

```
/************************************************************
File name   : f0206.cpp
Description :  构造函数的重载
************************************************************/
#include <iostream>
using namespace std;

class   Date
{
    int year;
    int month;
    int day;
public:
    Date()
    {
        year=2015;
        month=9;
        day=1;
    }
    Date(int y,int m,int d)
    {
        year=y;
        month=m;
```

```
            day=d;
        }
        void print()
        {
            cout<<year<<"-"<<month<<"-"<<day<<endl;
        }
    };
    void main()
    {
        Date oneday,today(2015,2,28),tomorrow(2015,3,1);
        oneday.print();
        today.print();
        tomorrow.print();
    }
    //=================================
```

程序的输出结果为：

```
2015-9-1
2015-2-28
2015-3-1
```

程序中，既可以用默认构造函数为没有提供参数的 oneday 赋值 2015,9,1，也可以利用带 3 个参数的构造函数，分别为 today 和 tomorrow 赋值。

创建多个构造函数的另一个目的是，可以更好地满足用户的需求。例如，对于日期的书写的格式，英国和美国是不同的。英国式表示法是日在前，月份在后；美国式表示法则与此相反。如 2015 年 3 月 2 日的写法：

```
2nd, March, 2015 (英)
March 2, 2015    (美)
```

同样，全部用数字表达日期时，英美也有差别。02.03.2015 是英国式的 2015 年 3 月 2 日，按照美国的表达方式却是 2015 年 2 月 3 日，美国的 2015 年 3 月 2 日应写成 03,02,2015。因此，全部使用数字来表示日期时，往往会产生误解。可以将月写成英文的简写形式：Jan,Feb,Mar 等。并且同时提供按 Mar,2,2015 和 2,Mar,2015 两种构造方法，这样就解决了英美在日期表达方面可能造成的误解，同时也满足了英美人的不同使用偏好。

```
    /***************************************************************
        File name   :  f0207.cpp
        Description :  构造函数的重载举例
    ***************************************************************/
    #include <iostream>
    using namespace std;

    enum Month
    {
        Jan=1,Feb,Mar,Apr,May,Jun,Jul,Aug,Sep,Oct,Nov,Dec
    };
    class   Date
    {
```

```cpp
        int year;
        Month month;
        int day;
public:
        inline Date(int y=2015,int m=9,int d=1);
        Date(int d,Month m,int y)
        {
            year=y;
            month=m;
            day=d;
        }
        Date(Month m,int d,int y)
        {
            year=y;
            month=m;
            day=d;
        }
        bool isLeapYear()
        {
            if( year%400==0 || (year%4==0 && year%100!=0))
                return true;
            else
                return false;
        }
        void init_date(int y,int m,int d)
        {
            year=y;
            month=Month(m);
            day=d;
        }
        void print()
        {
            cout<<year<<"-"<<month<<"-"<<day<<endl;
        }
};
Date::Date(int y,int m,int d)
{
    year=y;
    month=Month(m);
    day=d;
}
void main()
{
    Date oneday(Jan,28,2015);
    oneday.print();
    Date today(28,Feb,2015);
    today.print();
    Date tomorrow;
```

```
        if(today.isLeapYear())
            tomorrow.init_date(2015,2,29);
        else
            tomorrow.init_date(2015,3,1);
        tomorrow.print();
    }
//===================================
```

程序的输出结果为：

```
2015-1-28
2015-2-28
2015-3-1
```

该程序中，共有 3 个构造函数。Date(int d,Month m,int y)用于定义(日:月:年)顺序的对象；Date(Month m,int d,int y)用于定义(月:日:年)顺序的对象。由于定义了枚举型 Month，3 个参数的类型不同，可以重载。Date(int y=2015,int m=9,int d=1)是一个带有 3 个默认参数值的构造函数，以(年:月:日)序列定义对象，这里的 3 个参数都是整型量，也和上面 2 个构造函数类型不同，也可以重载。由于参数都带有默认值，所以可以构造无参对象。

有关构造函数的使用，有以下说明：

(1)在类对象进入其作用域时自动调用构造函数。因此，构造函数不需用户调用，也不能被用户调用。

(2)构造函数没有返回值，因此不需要也不能够在定义构造函数时声明类型，包括 void 也不行。

(3)如果用户自己没有定义构造函数，则 C++系统会自动生成一个构造函数，只是这个构造函数的函数体是空的，没有参数，也不执行任何操作，称为默认构造函数。

(4)一旦用户创建了自己的构造函数，系统将不创建默认构造函数了。此时，需要特别注意自己定义无参构造函数，或者带全部默认参数值的构造函数，否则无法创建无参的对象。

(5)如果在建立对象时选用的是无参构造函数，应注意正确书写定义对象的语句。请区别下面的语句：

```
    Date day1(2015,8,7);      //对
```

这里定义了 Date 类的一个对象 day1，并初始化为 2015 年 8 月 7 日。

```
    Date day2();              //错
```

这里声明了一个函数 day2，不带参数，返回值类型是 Date，没有函数定义。

如果需要定义一个不赋初值的对象，正确的方法应该是：

```
    Date day3;                //对
```

(6)尽管在一个类中可以包含多个构造函数，但是对于每一个对象来说，建立对象时只执行其中一个构造函数，并非每个构造函数都被执行。

(7)在构造函数中使用默认参数是方便有效的方法，它提供了建立对象时的多种选择，其作用相当于多个重载的构造函数。它的好处是，即使在调用构造函数时没有提供实参值，不仅不会出错，而且还确保按照默认的参数值将对象初始化。尤其在希望对每一个对象都有同样的初始化状况时，用这种方法更为方便。

关于构造函数默认值的几点说明：

(1) 应该在声明构造函数时指定默认值，而不能只在定义构造函数时指定默认值。

假如：将 f0207.cpp 中类的声明改为：

```
Date(int y,int m, int d);
```

类外的实现改成

```
Date::Date(int y=2015,int m=9, int d=1)
{
    year=y;
    month=m;
    day=d;
}
```

程序在 VC6.0 下，将无法通过编译。甚至，将类中的声明改为：

```
Date(int y=2015,int m=9, int d=1);
```

程序都无法通过编译。只有将类外的实现改为无默认参数值，类内的声明带有默认参数值时，才能编译通过。

(2) 如果构造函数形参带默认值，则必须从右到左依次带默认值，中间不能断开。可以：

```
Date::Date(int y=2015,int m=9, int d=3)
Date::Date(int y,int m=9, int d=3)
Date::Date(int y,int m, int d=3)
```

但是不允许：

```
Date::Date(int y=2015,int m, int d=3)
```

(3) 如果构造函数的全部参数都指定了默认值，则在定义对象时可以给一个或几个实参，也可以不给出实参。如果有实参，必须是从左到右都带实参，中间不能间断。

对于下面的构造函数：

```
Date::Date(int y=2015,int m=9, int d=3);
```

定义对象 date 可以：

```
Date date(2015,10,19);      //2015,10,19
Date date(2015,10);         //2015,10,3
Date date(2014);            //2014,9,3
Date date;                  //2015,9,3
```

但是不允许：

```
Date date(2015,,1);         //error
```

(4) 在一个类中定义了全部是默认参数的构造函数后，特别要注意，不要产生歧义。

```
/***************************************************************
File name    :  f0208.cpp
Description  :  构造函数重载的歧义问题
***************************************************************/
#include <iostream>
using namespace std;

enum Month
```

```cpp
{
    Jan=1,Feb,Mar,Apr,May,Jun,Jul,Aug,Sep,Oct,Nov,Dec
};
class   Date
{
    int year;
    Month month;
    int day;
public:
    Date(int y=2015,int m=9,int d=1)
    {
        year=y;
        month=Month(m);
        day=d;
    }
    Date(int d,Month m,int y)      //注意: Date(int d,Month m=Sep,int y=1)不可以
    {
        year=y;
        month=m;
        day=d;
    }
    Date(Month m,int d=1,int y=2015)
    {
        year=y;
        month=m;
        day=d;
    }
    bool isLeapYear()
    {
        if( year%400==0 || (year%4==0 && year%100!=0))
            return true;
        else
            return false;
    }
    void init_date(int y,int m,int d)
    {
        year=y;
        month=Month(m);
        day=d;
    }
    void print()
    {
        cout<<year<<"-"<<month<<"-"<<day<<endl;
    }
};
void main()
{
    Date oneday(Jan,28,2015);
```

```
        oneday.print();
        Date today(28,Feb,2015);
        today.print();
        Date tomorrow(2015);
        if(today.isLeapYear())
            tomorrow.init_date(2015,2,29);
        else
            tomorrow.init_date(2015,3,1);
        tomorrow.print();
    }
    //================================
```

程序的输出结果为：

```
    2015-1-28
    2015-2-28
    2015-3-1
```

该程序中，可以建立带 2 个默认参数值的构造函数 Date(Month m,int d=1,int y=2015)。因为该函数和带 3 个默认参数值的构造函数 Date(int y=2015,int m=9,int d=1)在第 1 个参数的类型上不同，所以在建立带 1 个参数的对象时可以相互区别：

```
    Date day(Jan);
    Date day(2015);
```

但是，如果建立 2 个默认参数值的构造函数 Date(int d,Month m=Sep,int y=1)时，却会产生歧义。因为该函数和带 3 个默认参数值的构造函数 Date(int y=2015,int m=9,int d=1)在第 1 个参数的类型相同，所以在建立带 1 个参数的对象时无法相互区别：

```
    Date day(1);
    Date day(2015);
```

对这类问题，在编写程序和调试程序时，都要特别注意。

2.3.4 参数初始化列表

上面介绍的是在构造函数的函数体内通过赋值语句对数据成员实现初始化。C++还提供另一种初始化数据成员的方法——参数初始化列表来实现对数据成员的初始化。这种方法不是在函数体内对数据成员初始化，而是在函数的首部实现。这样做的作用何在？在哪些情况下必须使用？

在 C 语言中，下面语句

```
    int i=5;           //初始化
```

在定义整型变量 i 的同时，将其初始化为 5。也可以写成：

```
    int i(5);          //等同于 int i=5;
```

而下面语句：

```
    int i;             //定义变量 i
    i=5;               //赋值
```

则是先定义了整型变量 i，然后再将其赋值为 5。

带初始化列表的构造函数的一般形式如下：

```
类名::构造函数名(<参数列表>)[:<成员初始化列表>]
{
    构造函数体
}
```

其中，尖括号内为可选项，可以出现，也可以不出现。

通过下面的例子可以了解怎样利用初始化列表。

```cpp
/******************************************************************
    File name    :   f0209.cpp
    Description  :   利用参数初始化列表构造对象
******************************************************************/
#include <iostream>
using namespace std;
enum Month
{
    Jan=1,Feb,Mar,Apr,May,Jun,Jul,Aug,Sep,Oct,Nov,Dec
};
class   Date
{
    int year;
    Month month;
    int day;
public:
    Date(int y=0,int m=0,int d=0):year(y),month(Month(m)),day(d)
    {}
    Date(int d,Month m,int y):year(y),month(m),day(d)
    {}
    Date(Month m,int d,int y):year(y),month(m),day(d)
    {}
    bool isLeapYear()
    {
        if( year%400==0 || (year%4==0 && year%100!=0))
            return true;
        else
            return false;
    }
    void init_date(int y,int m,int d)
    {
        year=y;
        month=Month(m);
        day=d;
    }
    void print()
    {
        cout<<year<<"-"<<month<<"-"<<day<<endl;
    }
};
```

```cpp
void main()
{
    Date today(Jan,28,2015);
    today.print();
    Date oneday(28,Feb,2015);
    oneday.print();
    Date tomorrow;
    if(today.isLeapYear())
        tomorrow.init_date(2015,2,29);
    else
        tomorrow.init_date(2015,3,1);
    tomorrow.print();
}
//=====================================
```

程序的输出结果为:

```
2015-1-28
2015-2-28
2015-3-1
```

有时候，一些成员的初始化必须使用初始化列表。这里涉及常量和引用，还有类对象成员也必须使用初始化列表进行构造和初始化。

1. 常量的初始化

对于常量，必须在定义的同时对其初始化:

```cpp
const int i=5;   //或者 const int i(5);，两条语句意义相同。
```

但是，不能定义常量，再赋值给它。

```cpp
const int i;    //对, i 的值为 0;
i=5;            //错
```

因为，该量是常量，不能改变其值。同样的道理，类中定义的常量成员不能在构造函数的函数体内赋值，必须通过初始化列表将其初始化。

```cpp
/*******************************************************************
File name   : f0210.cpp
Description : 常量成员的初始化
*******************************************************************/
#include <iostream>
using namespace std;
class Circle
{
    double radius;
    const double PI;
public:
    Circle(double r,double pi): radius(r),PI(pi){}
    double getArea()
    {
        return PI*radius*radius;
```

```
        }
    };

    void main()
    {
        Circle r1(1.1,3.14);
        cout<<"the area is "<<r1.getArea()<<endl;
        Circle r2(1.1,3.141593);
        cout<<"the area is "<<r2.getArea()<<endl;
    }
    //===================================
```

程序的输出结果为：

```
    the area is 3.7994
    the area is 3.80133
```

该程序中，类 Circle 定义了一个静态常量 PI，用以表示圆周率。在形成不同对象时，可以调整圆周率的精度。r1 对象的 PI 取值 3.14，r2 对象的 PI 取值 3.141593。由于 PI 是常量，不能放在函数体里面赋值：

```
    Circle(double r,double pi)
    {
        radius=r;           //可以，赋值语句
        PI=pi;              //不可以，因为 PI 是常量，不能赋值
    }
```

此时，只能使用初始化列表对其初始化。

2. 引用的初始化

引用也必须使用初始化列表进行初始化。例如：

```
    /*******************************************************************
    File name   : f0211.cpp
    Description : 引用的初始化
    *******************************************************************/
    #include <iostream>
    using namespace std;

    class A
    {
        int i;
        int &ri;
        int &rii;
    public:
        A(int ii):i(ii),ri(i),rii(ii)
        {
            cout<<"in constructor\n"<<" i= "<<i
            <<"   ri= "<<ri<<"   rii= "<<rii<<endl;
        }
        void print()
```

```
            {
                 cout<<" i= "<<i<<"    ri= "<<ri<<"    rii= "<<rii<<endl;;
            }
    };
    void main()
    {
        A a(2);
        a.print();
    }
    //=====================================
```

程序的输出结果为：

```
in constructor
i= 2     ri= 2     rii= 2
i= 2     ri= 2     rii= 4199850
```

这里，i 被初始化为 ii，值为 2，ri 通过初始化列表初始化为 i，所以 i 和 ri 值都为 2。而 rii 通过初始化列表初始化为 ii，由于 ii 是构造函数的形参，随着构造函数的运行结束而消失，因此后面的 rii 将是不确定的结果。这个例子告诉我们，定义引用时一定要确定被引用对象的生命周期，不要搞错。

3. 类对象成员的初始化

该部分的内容将在 3.4 节介绍。

2.3.5 析构函数

析构函数(destructor)也是一类特殊的成员函数，作用与构造函数相反。它的名字是类名之前加一个"~"符号，不带参数，也没有返回值。当对象的生命期结束时，会自动执行析构函数。

和默认构造函数一样，如果不自己定义一个析构函数，系统会为用户自动生成一个默认析构函数。该函数什么也不做。这对于一般的变量来讲，没有问题，因此前面的程序都没有提供自己定义的析构函数。但是当对象的成员变量从堆中申请空间后，则必须在析构函数中将其释放，否则该空间无法自动释放，可能造成内存泄漏。

```
/****************************************************************
File name    : f0212.cpp
Description  : 析构函数的实例
****************************************************************/
class Person
{
    char* pName;
public:
    Person(char* pN="noName")
    {
        cout<<"Constructing "<<pN<<"\n";
        pName = new char[strlen(pN)+1];
        if(pName)
```

```
            strcpy(pName,pN);
        }
        ~Person()
        {
            cout <<"Destructing "<<pName<<"\n";
            delete []pName;
        }
    };
    void main()
    {
        Person p1("Randy");
    }
    //=================================
```

这里，构造函数利用 new 申请了空间，则需要析构函数利用 delete 释放该空间，否则系统无法自行释放。

当出现以下几种情况时，程序会执行析构函数：

(1) 如果在一个函数中定义了一个对象，该对象是自动局部对象。当这个函数被调用结束时，在对象释放前自动执行析构函数。

(2) 静态局部对象在函数调用结束时对象并不释放，因此也不调用析构函数，只有在 main 函数结束或调用 exit 函数结束程序时，才调用静态局部对象的析构函数。

(3) 如果定义了一个全局对象，则在 main 函数结束或调用 exit 函数结束程序时，才调用该全局对象的析构函数。

(4) 如果用 new 运算符动态地建立了一个对象，当用 delete 运算符释放该对象之前，才调用该对象的析构函数。

析构函数的作用并不是删除对象，而是在撤销对象占用的各种资源前完成一些清理工作，使这部分内存可以被程序分配给新对象使用。程序设计者事先设计好析构函数，以完成所需的功能，只要对象的生命期结束，程序就会自动执行析构函数来完成这些工作。

特别要注意，析构函数不返回任何值，没有函数类型，也没有函数参数。因此它不能被重载。一个类可以有多个构造函数，但只能有一个析构函数。

2.4 对象的复制与赋值

2.4.1 复制构造函数

有时需要用到多个完全相同的对象，例如，同一型号的每一个产品从外表到内部属性都是一样的，如果要对每一个产品分别进行处理，就需要建立多个同样的对象，并要进行相同的初始化，用以前的办法定义对象比较麻烦。

C++提供了复制对象的方法，用来实现上述功能，这就是对象的复制机制。用一个已有的对象复制出多个完全相同的对象，实现该功能的函数称为复制构造函数(copy constructor)，或者复制构造函数。f0213.cpp 给出了复制构造函数使用的例子。

```
/****************************************************************
    File name : f0213.cpp
```

```
Description :  复制构造函数
****************************************************************/
#include <iostream>
using namespace std;
class CPoint
{
public:
        CPoint(int x = 0, int y = 0)
        {
            nPosX = x;
            nPosY = y;
        }
        CPoint(const CPoint& pt):nPosX(pt.nPosX),nPosY(pt.nPosY)
        {}
        ~CPoint()
        {}
        void print()
        {
            cout<<"X="<<nPosX<<",Y="<<nPosY<<endl;
        }
private:
        int nPosX, nPosY;
};

void main()
{
    CPoint pt1(1,2);
    pt1.print();
    CPoint pt2(pt1);
    pt2.print();
    CPoint pt3=pt2,pt4(pt3);
    pt3.print();
    pt4.print();
}
//====================================
```

程序的输出结果为：

```
X=1,Y=2
X=1,Y=2
X=1,Y=2
X=1,Y=2
```

运行后，4个对象的状态完全相同。

复制对象的语句

```
CPoint pt2(pt1);
```

利用 pt1 建立一个新对象 pt2。由于在括号内给定的实参是对象，因此编译系统就调用复制构造函数，不会去调用其他构造函数。实参 pt1 以引用的形式传递给形参 pt，因此执行复制构造函数的函数体时，将 pt1 对象中各数据成员的值赋给 pt2 中各数据成员。

复制构造函数的一般形式为：

```
类名  对象2(对象1);
```

用对象 1 复制出对象 2。

可以看到，它与定义对象的方式类似，但是括号中给出的参数不是一般的变量，而是对象的引用：

```
CPoint(const CPoint& pt)
{
    nPosX = pt.nPosX;
    nPosY = pt.nPosY;
}
```

或者用初始化列表来实现：

```
CPoint(const CPoint& pt):nPosX(pt.nPosX),nPosY(pt.nPosY)
{}
```

复制构造函数也是构造函数，它只有一个参数，该参数是本类的对象，而且采用对象的引用形式，一般约定加 const 声明，使参数值不能改变，以免在调用此函数时因不慎而修改对象值。此复制构造函数的作用就是将实参对象的各成员值一一赋给新的对象中对应的成员。

如果用户自己未定义复制构造函数，则编译系统会自动提供一个默认的复制构造函数，其作用只是简单地复制类中每个数据成员。

C++还提供另一种方便用户的复制形式，用赋值号代替括号，如

```
CPoint pt2=pt1;         //用 pt1 初始化 pt2
```

其一般形式为

```
类名 对象名1 = 对象名2;
```

可以在一个语句中进行多个对象的复制。如

```
CPoint pt2=pt1,pt3=pt2;
```

用 pt1 来复制 pt2，并用 pt2 复制 pt3。这种形式看起来很直观，用起来很方便，其作用都是调用复制构造函数。

注意：对象的复制和对象的赋值在概念上和语法上不同。对象的赋值是对一个已经存在的对象赋值，因此必须先定义被赋值的对象，才能进行赋值。而对象的复制则是从无到有地建立一个新对象，并使它与一个已有的对象完全相同。关于赋值，将在 2.4.3 节中详述。

普通构造函数和复制构造函数的区别主要有以下几个方面：

(1)在形式上。

```
类名(形参表列);              //普通构造函数的声明，如 CPoint(int x,int y);
类名(const 类名&对象名);      //复制构造函数的声明，如 CPoint(const CPoint &pt);
```

(2)在建立对象时，实参类型不同。

系统会根据实参的类型决定调用普通构造函数或者复制构造函数。如

```
CPoint pt1(1,2);            //实参为 2 个整数，调用普通构造函数
CPoint pt2(pt1);            //实参是对象名，调用复制构造函数
```

(3) 在什么情况下被调用。

普通构造函数在程序建立对象时被调用。复制构造函数在用已有对象复制一个新对象时被调用，在以下 3 种情况下需要复制对象：

①程序中需要新建一个对象，并用另一个同类的对象对它初始化。例如：

```
CPoint pt2(pt1);
```

②当函数的参数为类的对象时。在调用函数时需要将实参对象完整地传递给形参，也就是需要建立一个实参的复制，这就是按实参复制一个形参，系统是通过调用复制构造函数来实现的，这样能保证形参具有和实参完全相同的值。例如：

```
/**********************************************************************
    File name   : f0214.cpp
    Description : 函数参数的值传递
**********************************************************************/
#include <iostream>
using namespace std;
class CPoint
{
public:
    CPoint(int x = 0, int y = 0)
    {
        cout<<"in constructor"<<endl;
        nPosX = x;
        nPosY = y;
    }
    CPoint(const CPoint& pt):nPosX(pt.nPosX),nPosY(pt.nPosY)
    {
        cout<<"in copy constructor"<<endl;
    }
    ~CPoint()
    {
        cout<<"in destructor   ";
        cout<<"X="<<nPosX<<",Y="<<nPosY<<endl;
    }
    void add()
    {
        nPosX++;
        nPosY++;
    }
private:
    int nPosX, nPosY;
};
void fun(CPoint pt)        //形参是类的对象
{
    pt.add();
}
```

```
void main()
{
    CPoint pt1(1,2);
    fun(pt1);      //实参是类的对象,调用函数时将复制一个新对象
    return;
}
//=================================
```

程序的输出结果为:

```
in constructor
in copy constructor
in destructor    X=2,Y=3
in destructor    X=1,Y=2
```

这里,CPoint pt1(1,2)语句调用构造函数,fun(pt1)调用复制构造函数。当函数 fun 运行结束,系统自动调用析构函数将形参对象释放掉,显示的是 X=2,Y=3;当 main 函数结束时,系统自动运行析构函数,将 pt1 释放掉。

引用和指针作为参数是不形成新对象的。请读者运行下列函数,分析运行结果。

```
void fun(CPoint &pt)       //形参是类对象的引用
{}
void fun(CPoint* pt)       //形参是类对象的指针
{}
```

③函数的返回值是类的对象。当函数调用完毕把返回值带回函数调用处时,函数中的对象调用复制构造函数,复制一个临时对象并传给该函数的调用处。例如:

```
/******************************************************************
    File name    :  f0215.cpp
    Description  :  函数返回对象的值传递
******************************************************************/
#include <iostream>
using namespace std;
class CPoint
{
public:
        CPoint(int x = 0, int y = 0)
        {
            nPosX = x;
            nPosY = y;
            cout<<"in constructor  ";
            cout<<"X="<<nPosX<<",Y="<<nPosY<<endl;
        }
        CPoint(const CPoint& pt):nPosX(pt.nPosX),nPosY(pt.nPosY)
        {
            cout<<"in copy constructor  ";
            cout<<"X="<<nPosX<<",Y="<<nPosY<<endl;
        }
```

```cpp
            ~CPoint()
            {
                cout<<"in destructor   ";
                cout<<"X="<<nPosX<<",Y="<<nPosY<<endl;
            }
            void add()
            {
                nPosX++;
                nPosY++;
            }
    private:
            int nPosX, nPosY;
    };
    CPoint f()              //函数 f 的类型为 CPoint 类类型
    {
        CPoint pt1(1,5);
        return pt1;         //返回值是 CPoint 类的对象
    }
    void main()
    {
        CPoint pt2;         //定义 CPoint 类的对象 pt2
        pt2=f();            //调用 f 函数,返回 CPoint 类的临时对象,并将它赋值给 pt2
    }
    //==================================
```

程序的输出结果为:

```
    in constructor   X=0,Y=0
    in constructor   X=1,Y=5
    in copy constructor   X=1,Y=5
    in destructor    X=1,Y=5
    in destructor    X=1,Y=5
    in destructor    X=1,Y=5
```

将主函数改为:

```cpp
    void main()
    {
        CPoint pt2(f());    //利用 f()返回值构造 CPoint 类的对象 pt2
        pt2.add();
    }
    //==================================
```

程序的输出结果为:

```
    in constructor   X=1,Y=5
    in copy constructor   X=1,Y=5
    in destructor    X=1,Y=5
    in destructor    X=2,Y=6
```

2.4.2 深复制和复制构造函数

在某些状况下，类内成员变量需要动态开辟堆内存，如果实行所谓的逐位复制，也就是把对象 B 里的值完全复制给另一个对象 A，这时，如果 B 中有一个成员变量指针已经申请了内存，那 A 中的那个成员变量也指向同一块内存。这就出现了问题：当 B 通过析构函数释放内存，这时 A 的指针就成为野指针，一旦 A 对象再运行析构函数，将出现运行错误。

```cpp
/***************************************************************
    File name    : f0216.cpp
    Description  : 浅复制存在的问题
***************************************************************/
#include <iostream>
#include <cstring>
using namespace std;

class CName
{
public:
    CName()
    {
        strName = NULL;                         //空值
        cout<<"默认构造函数！"<<endl;
    }
    CName( char *str )
    {
        strName = new char[strlen(str)+1];
        strcpy( strName, str );                 //复制内容
        cout<<"构造函数！"<<endl;
    }
    ~CName()
    {
        if (strName)
            delete []strName;
        strName = NULL;                         //一个好习惯
        cout<<"析构函数！"<<endl;
    }
    char *getName()
    {
        return strName;
    }
    void setName( char *str )
    {
        if (strName)
            delete []strName;
        strName = new char[strlen(str)+1];
        strcpy( strName, str );                 //复制内容
    }
```

```
    private:
        char    *strName;                          //字符指针
};
void print( CName one)
{
    cout<<one.getName()<<endl;
}

void main()
{
    CName person1("qu weiguang");
    CName person2(person1);
//    print(person1);
//    print(person2);
}
```

程序的输出结果为:

```
构造函数!
析构函数!
```

当我们把注释掉的 2 条语句加入主函数,运行结果为:

```
构造函数!
qu weiguang
析构函数!
茸茸茸茸茸茸茸茸?
```

这时就必须提供深复制的复制构造函数:

```
CName( const CName &one )                     //复制构造函数
    {
        //为 strName 开辟独立的内存空间
        strName = new char[strlen(one.strName)+1];
        strcpy(strName, one.strName);              //复制内容
        cout<<"复制构造函数!"<<endl;
    }
```

程序的运行结果为:

```
构造函数!
复制构造函数!
复制构造函数!
```

qu weiguang

```
析构函数!
复制构造函数!
```

qu weiguang

```
析构函数!
析构函数!
析构函数!
```

深复制和浅复制可以简单理解为:如果一个类拥有资源,当这个类的对象发生复制过程

时，资源重新分配，这个过程就是深复制，反之，没有重新分配资源，就是浅复制。

如果在类中没有显式地声明一个复制构造函数，那么编译器将会自动生成一个默认的复制构造函数，该构造函数完成对象之间的浅复制。如果有资源重新分配，如使用 new 申请了堆空间，则必须自定义复制构造函数，这就是所谓的深复制。

2.4.3 对象的赋值

在学习 C 语言时，下面的语句：

```
int i=5;
int j=i;    //利用 i 初始化 j;
```

先定义变量 i，并通过 i 来实现对 j 的初始化。学习了 C++，可以通过：

```
CPoint pt1(1,2);
CPoint pt2(pt1);
```

来实现利用 pt1 实现对 pt2 的初始化。

在学习 C 语言时，下面的语句，可以实现用 i 对 j 赋值：

```
int i=5;
int j;
j=i;    //利用 i 赋值 j;
```

甚至可以

```
i=i;    //自己给自己赋值;
```

同样，对一个类定义了两个或多个对象，则这些同类的对象之间可以互相赋值，或者说，一个对象的值可以赋给另一个同类的对象，甚至可以允许自己赋值给自己。这里所指对象的值是指对象中所有数据成员的值。

对象之间的赋值也是通过赋值运算符 "=" 进行的。本来，赋值运算符 "=" 只能用来对内嵌数据类型的变量赋值，现在被扩展为两个同类对象之间的赋值，这是通过对赋值运算符的重载实现的。

这个过程是通过成员复制来完成的，即将一个对象成员的值一一复制给另一对象的对应成员。

对象赋值的一般形式为：

```
对象名1 = 对象名2;
```

注意对象名 1 和对象名 2 必须属于同一个类。例如

```
Box box1(3,4,5),box2;    //定义两个同类的对象
bos2=box1;               //将 box1 赋给 box2
```

通过下面的例子可以了解怎样进行对象的赋值。

```
/*************************************************************
    File name    : f0217.cpp
    Description  : 对象的赋值
*************************************************************/
#include<iostream>
using namespace std;
```

```cpp
class Box
{
public:
    Box(int=10,int=10,int=10);          //声明有默认参数的构造函数
    int volume();
private:
    int height;
    int width;
    int length;
};
Box::Box(int h,int w,int len)
{
   height=h;
   width=w;
   length=len;
}
int Box::volume()
{
   return(height*width*length);          //返回体积
}
int main()
{
   Box box1(1,6,4),box2;                 //定义两个对象box1和box2
   cout<<"The volume of box1 is "<<box1.volume()<<endl;
   box2=box1;                            //将box1的值赋给box2
   cout<<"The volume of box2 is "<<box2.volume()<<endl;
   return 0;
}
```

程序的输出结果为:

```
The volume of box1 is 24
The volume of box2 is 24
```

这里并没有提供任何函数实现赋值运算(=)。同默认复制构造函数一样，编译系统提供了默认的赋值函数，可以提供类似于默认复制构造函数浅复制的功能。

对象的赋值只对其中的数据成员赋值，而不对成员函数赋值。数据成员是占存储空间的，不同对象的数据成员占有不同的存储空间，赋值的过程是将一个对象的数据成员在存储空间的状态复制给另一对象的数据成员的存储空间。而不同对象的成员函数是同一个函数代码段，不需要、也无法对它们赋值。

类的数据成员中不能包括动态分配的数据，否则在赋值时也会出现类似深复制和浅复制的问题，可能出现严重后果。

```cpp
/***************************************************************
    File name    :   f0218.cpp
    Description  :   对象的赋值
***************************************************************/
#include <iostream>
using namespace std;
```

```cpp
class Person
{
    char* pName;
public:
    Person(char* pN="noName")
    {
        pName = new char[strlen(pN)+1];
        if(pName)
            strcpy(pName,pN);
        cout<<"构造函数: "<<pName<<endl;
    }
    Person(const Person& s)
    {
        pName = new char[strlen(s.pName)+1];
        if(pName)
            strcpy(pName, s.pName);
        cout<<"复制构造函数: "<<pName<<endl;
    }
    ~Person()
    {
        cout<<"析构函数: "<<pName<<endl;
        delete[] pName;
    }
    void print()
    {
        cout<<pName<<"    \n";
    }
};

void main()
{
    Person a1("zhang"),a2="wang";
    a1.print();
    a2.print();
    Person temp;
    temp.print();
    temp=a1;
    temp.print();
}
```

程序的输出结果为:

```
构造函数: zhang
构造函数: wang
zhang
wang
构造函数: noName
noName
zhang
```

```
析构函数： zhang
析构函数： wang
析构函数： 茸茸茸茸茸茸茸茸_
```

赋值操作(=)实际是一个函数，temp=a1 的函数形式类似于带两个参数的函数=(temp,a1)。就像深复制和浅复制问题，用户也必须将等号赋值函数进行重新定义。定义运算符将在第 4 章运算符重载中讲解，这里先定义一个赋值函数 Assign：

```
Person& Assign(const Person & a)
{
    if(strcmp((*this).pName,a.pName)!=0)
    {
        delete []pName;
        pName=new char[strlen(a.pName)+1];
        strcpy(pName,a.pName);
    }
    return *this;
}
```

该函数等价于重载的运算符=函数：

```
Person& operator =(const Person & a)
{
    if(strcmp((*this).pName,a.pName)!=0)
    {
        delete []pName;
        pName=new char[strlen(a.pName)+1];
        strcpy(pName,a.pName);
    }
    return *this;
}
```

将其加入类中成为 public 成员函数，并运行之，结果为：

```
构造函数：zhang
构造函数：wang
zhang
wang
构造函数：noName
noName
zhang
析构函数：zhang
析构函数：wang
析构函数：zhang
```

函数中用到了 this 指针。this 指针是一个隐含于每一个成员函数中的特殊指针。它是一个指向正在被该成员函数操作的对象，也就是要操作该成员函数的对象。

当对一个对象调用成员函数时，编译程序先将对象的地址赋给 this 指针，然后调用成员函数，每次成员函数存取数据成员时，有隐含作用 this 指针。通常不去显式地使用 this 指针来引用数据成员，但是也可以 this->这样来使用成员。当然，也可以使用*this 来标识调用该成员函数的对象。

2.5 构造函数和析构函数调用的顺序

在使用构造函数和析构函数时，需要特别注意对它们的调用时间和调用顺序。在一般情况下，调用析构函数的次序正好与调用构造函数的次序相反：最先被调用的构造函数，其对应的析构函数最后被调用，而最后被调用的构造函数，其对应的析构函数最先被调用。

可以简记为：先构造的后析构，后构造的先析构，它相当于一个栈，后进先出。

但是，并不是在任何情况下都是按这一原则处理的。对象的作用域和存储类别对于对象也会产生影响。对象可以在不同的作用域中定义，可以有不同的存储类别，这些会影响调用构造函数和析构函数的时机。

什么时候调用构造函数和析构函数的情况如下：

(1)在全局范围中定义的对象(即在所有函数之外定义的对象)，它的构造函数在文件中的所有函数(包括 main 函数)执行之前调用。但如果一个程序中有多个文件，而不同的文件中都定义了全局对象，则这些对象的构造函数的执行顺序是不确定的。当 main 函数执行完毕或调用 exit 函数终止程序运行时，调用析构函数。

(2)如果定义的是局部自动对象，则在建立对象时调用其构造函数。如果函数被多次调用，则在每次建立对象时都要调用构造函数。在函数调用结束、对象释放时先调用析构函数。

(3)如果在函数中定义静态局部对象，则只在程序第一次调用此函数建立对象时调用构造函数一次，调用结束时对象并不释放，因此也不调用析构函数，只在 main 函数结束或调用 exit 函数终止程序运行时，才调用析构函数。

```cpp
/****************************************************************
    File name     : f0219.cpp
    Description   : 构造函数和析构函数的调用顺序
****************************************************************/
#include <iostream>
#include <string>
using namespace std;

class Student
{
    string id;          //定义学号
    char* pName;        //定义姓名
    char gender;        //定义性别

public:
    Student(string id="1320103001", char* pN="noName",
    char gender='m' ) :id(id),gender(gender)
    {
        pName = new char[strlen(pN)+1];
        if(pName)
            strcpy(pName,pN);
        cout<<"构造函数: "<<id<<endl;
    }
```

```cpp
        Student(const Student& s)
        {
            id=s.id;
            pName = new char[strlen(s.pName)+1];
            if(pName)
                strcpy(pName, s.pName);
            gender=s.gender;
            cout<<"复制构造函数: "<<id<<endl;
        }
        ~Student()
        {
            cout<<"析构函数: "<<id<<endl;
            delete[] pName;
        }
        void print()
        {
            cout<<id<<"    "<<pName<<"    "<<gender<<endl;
        }
};
Student stu1("1320103001","qwg",'m');
void fn()
{
    static Student stu2("1320103002","yws",'m');
}
void main()
{
    Student *pStu=new Student[2];
    Student stu3("1320103005","sq",'f');
    fn();
    Student stu4("1320103006","wqq",'f');
    fn();
    delete []pStu;
}
//====================================
```

程序的输出结果为:

```
构造函数: 1320103001
in main()
构造函数: 1320103000
构造函数: 1320103000
构造函数: 1320103005
构造函数: 1320103002
构造函数: 1320103006
析构函数: 1320103000
析构函数: 1320103000
out main()
析构函数: 1320103006
析构函数: 1320103005
析构函数: 1320103002
```

程序运行前，首先创建全局对象 stu1，运行构造函数，输出构造函数：1320103001；然后进入主函数，输出 in main()。通过 new Student[2]在堆区创建两个对象的数组，调用默认参数的构造函数，输出两次构造函数：1320103000；创建对象 stu3，输出构造函数：1320103005；调用函数 fn()，创建静态对象 stu2，输出构造函数：1320103002；静态对象在调用结束时对象并不释放，因此也不调用析构函数；创建对象 stu4，输出构造函数：1320103006；调用函数 fn()，由于静态函数只创建一次，所以没有再创建对象，同样也没有释放对象。通过 delete []pStu 语句释放由 new Student[2]在堆区创建两个对象的数组，输出 2 次析构函数：1320103000。输出 out main()，main 函数运行结束，系统自动调用析构函数，将对象 stu4 和 stu3 释放。可以看出，析构的顺序与构造的顺序正好相反。然后再释放掉静态变量，最后释放全局变量。

练 习

1. 创建一个时间类 Time，拥有时、分和秒 3 个数据成员，成员函数包括构造函数可以设置任意时、分和秒的值，Add 函数可以为秒的值加 1，Print 函数可以打印当前对象的值。
2. 分析现有字符串处理的功能，建立一个能够对字符串进行处理的类 MyString。要求写出成员函数的功能，并实现之。
3. 设计一个矩形类 Rectangle，其数据成员包括矩形的左下角和右上角的坐标，要求该类具有调整矩形左上角和右下角两个坐标的函数，可以求矩形的面积。

第 3 章　类和对象的使用

面向对象的程序设计方法，通过建立类把事物的属性以及处理事物的方法封装到一起，为大型软件的设计提供便利。本章将讨论类和对象在各种程序设计环境中的应用，重点包括动态对象的创建和释放、对象数组和对象数组类、静态成员和友元的相关知识。

- **知识要点**
  学习动态对象的创建方法；
  学习静态成员、友元的实现方法；
  掌握类在各种程序环境中的使用方法。
- **探索思考**
  深复制与浅复制。
- **预备知识**
  复习类的声明方法，对象的定义和使用；
  构造函数和析构函数。

3.1　动态对象的创建与释放

使用类名定义的对象本质上都是静态的。在程序运行过程中，对象所占的空间不能随时释放。例如，定义一个对象：

```
Date today(Jan,1,2016);
```

或者定义一个对象数组：

```
Date days[365];
```

这些对象在各自的生命周期内无法释放，虽然后续程序可能不再需要，也只能等到该对象生命周期结束后才能释放掉。人们希望在需要用到对象时才建立对象，在不需要使用该对象时就撤销，释放其所占的内存空间以供其他数据使用，这样可以提高内存空间的利用率。

在 C++中，可以使用 new 运算符动态地分配内存，用 delete 运算符释放这些内存空间。这也适用于对象，用 new 运算符动态建立对象，用 delete 运算符撤销对象。

如果已经定义了一个 Date 类，可以用下面的方法动态建立一个对象：

```
new Date
```

编译系统在堆区开辟一段内存空间，并在此内存空间中存放一个 Date 类对象，同时调用该类的构造函数，以使该对象初始化。用 new 运算符动态地分配内存后，将返回一个指向该对象的指针的值，即所分配的内存空间的起始地址。用户可以获得这个地址，并通过这个地址来访问这个对象。

需要定义一个指向本类的对象的指针变量来存放该地址。如：

```
Date *pt;        //定义一个指向 Date 类对象的指针变量 pt
pt=new Date;     //在 pt 中存放了新建对象的起始地址
```

在程序中就可以通过 pt 访问这个新建的对象。如：

```
pt->init_date(2016,1,1);        //或者(*pt).init_date(2016,1,1);
```

C++还允许在执行 new 时，对新建立的对象进行初始化。如：

```
Date *pt=new Date(2016,1,1);
```

这种写法是把上面两个语句合并成一个语句，并指定初值。新对象中的 year、month 和 day 分别获得初值 2016、Jan（通过类型转换，由整型量转换成枚举型量）和 1。

在执行 new 运算时，如果内存量不足，无法开辟所需的内存空间。目前大多数 C++编译系统都令 new 返回一个 0 指针值，只要检测返回值是否为 0，就可判断分配内存是否成功。

当不再需要使用由 new 建立的对象时，可以用 delete 运算符予以释放。如：

```
delete pt;             //释放 pt 指向的内存空间，用来撤销 pt 指向的对象
```

此后程序不能再使用该对象，但该指针还存在，仍然可以随时通过赋值语句接收 Date 类型通过 new 申请到的空间的指针。

如果用一个指针变量 pt 先后指向不同的动态对象，应注意指针变量的当前指向，以免删错了对象。一个指针变量 pt 指向另一个动态对象之前，一定要先进行 delete 操作。执行 delete 运算符时，系统在释放内存空间之前会自动调用析构函数，完成有关善后清理工作。

例如：

```
Date *pt;                  //定义一个指向 Date 类对象的指针变量 pt
pt=new Date;               //在 pt 中存放了新建对象的起始地址
pt=new Date(2016,1,1);     //存在内存泄漏
```

此时，pt 的内容由 new Date 申请的空间地址被修改为 new Date(2016,1,1) 申请的空间地址，由于 new Date 申请的空间地址被丢掉，因而将无法释放。因此在申请 Date(2016,1,1) 之前，必须把 Date 申请的空间，利用 delete 释放掉。同样，Date(2016,1,1) 使用完毕，也要释放掉。上段代码需要修改成：

```
Date *pt;                  //定义一个指向 Date 类对象的指针变量 pt
pt=new Date;               //在 pt 中存放了新建对象的起始地址
…
delete pt;
pt=new Date(2016,1,1);     //不存在内存泄漏
…
delete pt;
```

在 C 语言中，堆内存的申请和释放，分别使用 malloc 和 free。但是利用 malloc 和 free 只能获得申请到的空间或者释放之，无法自动调用构造函数和析构函数，因此它们不支持面向对象编程。

```
/************************************************************
File name   : f0301.cpp
Description : 类对象的动态创建与释放
************************************************************/
#include <iostream>
using namespace std;
```

```cpp
enum Month
{
    Jan=1,Feb,Mar,Apr,May,Jun,Jul,Aug,Sep,Oct,Nov,Dec
};
class   Date
{
    int year;
    Month month;
    int day;
public:
    Date(int y=0,int m=0,int d=0):year(y),month(Month(m)),day(d)
    {
        cout<<"构造函数1"<<endl;
    }
    Date(int d,Month m,int y):year(y),month(m),day(d)
    {
        cout<<"构造函数2"<<endl;
    }
    Date(Month m,int d,int y):year(y),month(m),day(d)
    {
        cout<<"构造函数3"<<endl;
    }
    ~Date()
    {
        cout<<"析构函数"<<endl;
    }
    bool isLeapYear()
    {
        if( year%400==0 || (year%4==0 && year%100!=0))
            return true;
        else
            return false;
    }
    void init_date(int y,int m,int d)
    {
        year=y;
        month=Month(m);
        day=d;
    }
    void print()
    {
        cout<<year<<"-"<<month<<"-"<<day<<endl;
    }
};

void main()
{
```

```
        Date *pt1=new Date(2016,1,1);
        pt1->print();
        Date *pt2=(Date*) malloc(sizeof(Date));
        pt2->print();
        pt2->init_date(2016,1,1);
        pt2->print();
        delete pt1;
        free(pt2);
    }
    //====================================
```

程序的输出结果为:

```
构造函数1
2015-11-11
-842150451--842150451--842150451
2016-1-1
析构函数
```

主函数首先通过运算符 new 调用"构造函数 1"创建对象,将申请空间的首地址作为 pt1 的初始化值,print 函数显示的结果为 2016-1-1。利用库函数 malloc 申请 Date 空间,将申请空间的首地址作为 pt2 的初始化值,结果显示,没有构造函数调用,print 函数显示的结果为随机值。只有通过 init_date 函数对该指针指向的对象赋值。执行 delete 语句调用析构函数将 pt1 对应的对象释放掉,输出 2 行"析构函数",而 free(pt2) 则没有执行析构函数。可以看出,利用 new 和 delete 可以自动调用构造函数和析构函数,而利用 malloc 和 free 则不能自动调用构造函数和析构函数。因此,在面向对象程序设计时,要逐渐舍弃过去 C 语言的一些使用,充分发挥 C++支持面向对象程序设计的强大功能。

3.2 对象数组

数组不仅可以由简单变量组成,如整型数组的每一个元素都是整型变量;也可以由对象组成,对象数组的每一个元素都是同类的对象。

在日常生活中,有许多实体的属性是共同的,只是属性的具体内容不同。例如,班级的学生,每个学生的属性包括姓名、性别、年龄等。除此之外,还必须要有身份证号码或者学生证号码,否则,将可能有多个对象因为自然属性相同而无法区分。如果为每一个学生建立一个对象,需要许多个对象名,用程序处理很不方便。这时可以定义一个"学生类"对象数组,每一个数组元素是一个"学生类"对象。例如:

```
    Student stud[10];  //假设已声明了 Student 类,定义 stud 数组,有 10 个元素
```

在建立数组时,同样要调用构造函数。有 10 个元素,就需要调用 10 次构造函数。

在需要时可以在定义数组时提供实参以实现初始化,在花括号中分别写出构造函数并指定实参即可。

```
    /*****************************************************************
        File name    :  f0302.cpp
        Description  :  创建对象数组
```

```cpp
*******************************************************************/
#include <iostream>
#include <string>
using namespace std;
class Student
{
    string id;          //定义学号
    char* pName;        //定义姓名
    char gender;        //定义性别

public:
    Student(string id="1320103001", char* pN="noName",
    char gender='m' ) :id(id),gender(gender)
    {
        pName = new char[strlen(pN)+1];
        if(pName)
            strcpy(pName,pN);
    }
    Student(const Student& s)
    {
        id=s.id;
        pName = new char[strlen(s.pName)+1];
        if(pName)
            strcpy(pName, s.pName);
        gender=s.gender;
    }
    ~Student()
    {
        delete[] pName;
    }
    void print()
    {
        cout<<id<<"   "<<pName<<"   "<<gender<<endl;
    }
};
void main()
{
    Student stu[4]={Student("1320103001","qwg",'m'),
        Student("1320103002","yws",'m'),
        Student("1320103003","wdb",'m'),
        Student("1320103004","hy",'m')};
    for(int i=0;i<4;i++)
    {
        stu[i].print();
    }
}
//====================================
```

程序的输出结果为：

```
1320103001    qwg    m
1320103002    yws    m
1320103003    wdb    m
1320103004    hy     m
```

这里，**Student** 类中定义了学生的学号、姓名和性别。

如果构造函数有 3 个参数，分别代表学号、姓名、性别。则可以这样定义对象数组：

```
Student stu[4]={Student ("1320103001","qwg",'m'),
                Student ("1320103002","yws",'m'),
                Student ("1320103003","wdb",'m'),
                Student ("1320103004","hy",'m')};
```

在建立对象数组时，分别调用构造函数，对每个元素初始化。每一个元素的实参分别用括号括起来，对应构造函数的一组形参，不会混淆。

当需要输入更多的对象时，这种利用花括号，通过构造函数进行初始化的方式并不可取。最常用的方法是利用文件系统输入信息，并且根据每次输入学生的人数，自动创建数组。

假如已经创建了文件 stulist.txt，内容如下：

```
4
1320103001  qwg  m
1320103002  yws  m
1320103003  wdb  m
1320103004  hy   m
```

其中，第 1 行的数字表示学生人数。于是可以重新编写主函数如下：

```cpp
void main()
{
    ifstream in("stulist.txt");
    if (!in)
    {
        cout <<"Can't open the file!"<< endl;
        exit(0);
    }
    int num;
    in>>num;
    Student *per=new Student[num];
    string id;
    char name[20];
    char gen;
    int year,month,day;
    int i;
    for(i=0;i<num;i++)
    {
        in>>id>>name>>gen;
        per[i].setData(id,name,gen);
    }
    for(i=0;i<num;i++)
```

```
        {
            per[i].print();
        }
}
```

由于 Student *per=new Student[num];只能同时构造 num 个对象，无法分别初始化其各自的属性，所以必须使用默认构造函数，或者带全部默认参数的构造函数，然后通过成员函数为每个对象赋值。这里定义函数：

```
void setData(string id,char *p,char gen)
    {
        this->id=id;
        if(pName!=0)
            delete []pName;
        pName = new char[strlen(p)+1];
        if(pName)
            strcpy(pName, p);
        gender=gen;
    }
```

来实现赋值的任务。

同时可以看到，每次通过文件建立 Student 数组，都需要在主函数中完成，并且每次的过程几乎相同。是否能够建立一个类，可以实现对 Student 数组的创建和处理呢？下一节将讨论如何建立一个数组类，来实现对 Student 数组的整体操作。

3.3 建立数组类

用户可以建立这样一个类，它的成员由 Student 的数组构成，并且可以完成对数组成员的各种操作。

```
/******************************************************************
    File name    : f0303.cpp
    Description  : 建立数组类
******************************************************************/

#include <iostream>
#include <fstream>
#include <string>
using namespace std;
class Student
{
    string id;
    char* pName;
    char gender;
public:
    Student(string id="1320103001",char* pN="noName",
        char gender='m'): id(id), gender(gender)
    {
```

```cpp
        cout<<"in Student constructor\n";
        pName = new char[strlen(pN)+1];
        if(pName)
            strcpy(pName,pN);
    }
    Student(const Student& s)
    {
        cout<<"in Student copy constructor\n";
        id=s.id;
        pName = new char[strlen(s.pName)+1];
        if(pName)
            strcpy(pName, s.pName);
        gender=s.gender;
    }
    void setData(const Student& s)
    {
        if(this == &s)
            return;
        id=s.id;
        pName = new char[strlen(s.pName)+1];
        if(pName)
            strcpy(pName, s.pName);
        gender=s.gender;
    }
    void setData(string id,char *p,char gen)
    {
        this->id=id;
        pName = new char[strlen(p)+1];
        if(pName)
            strcpy(pName, p);
        gender=gen;
    }
    ~Student()
    {
        cout<<"in Student destructor\n";
        delete[] pName;
    }
    void print()
    {
        cout<<id<<"   "<<pName<<"   "<<gender<<endl;
    }
};
class CDyArray
{
public:
    CDyArray(int n=100)
    {
        cout<<"in CDyArray contstuctor"<<endl;
        ptrArray=new Student[n];
```

```cpp
        Max_size=n;
        Cur_size=0;
    }
    CDyArray(CDyArray& arr)
    {
        cout<<"in CDyArray copy contstuctor"<<endl;
        ptrArray=new Student[arr.Max_size];
        Max_size=arr.Max_size;
        Cur_size=arr.Cur_size;
        for(int i=0;i<Cur_size;i++)
        {
            ptrArray[i].setData(arr.ptrArray[i]);
        }
    }
    void copyit(CDyArray& arr)
    {
        Cur_size=arr.Cur_size;
        for(int i=0;i<Cur_size;i++)
        {
            ptrArray[i].setData(arr.ptrArray[i]);
        }
    }
    void appendit(CDyArray& arr)
    {
        for(int i=0;i<arr.Cur_size;i++)
        {
            ptrArray[Cur_size+i].setData(arr.ptrArray[i]);
        }
        Cur_size=Cur_size+arr.Cur_size;
    }
    void setData(Student *arr,int num)
    {
        for(int i=0;i<num;i++)
        {
            ptrArray[i].setData(arr[i]);
        }
        Cur_size=num;
    }
    void setData(Student arr,int i)
    {
        ptrArray[i].setData(arr);
        Cur_size++;
    }
    void print()
    {
        for(int i=0;i<Cur_size;i++)
            ptrArray[i].print();
    }
    ~CDyArray()
    {
```

```cpp
            cout<<"in CDyArray destructor"<<endl;
            delete []ptrArray;
        }
    private:
        int Max_size;
        int Cur_size;
        Student *ptrArray;
};

void main()
{
    ifstream in("stulist.txt");
    if (!in)
    {
        cout <<"Can't open the file!"<< endl;
        exit(0);
    }
    int num;
    in>>num;
    CDyArray stu(num);
    string id;
    char name[20];
    char gen;
    for(int i=0;i<num;i++)
    {
        in>>id>>name>>gen;
        stu.setData(Student(id,name,gen),i);
    }
    stu.print();
}
```

程序的输出结果为：

```
in CDyArray contstuctor
in Student constructor
in Student constructor
in Student constructor
in Student constructor
in Student constructor
in Student copy constructor
in Student destructor
in Student destructor
in Student constructor
in Student copy constructor
in Student destructor
in Student destructor
in Student constructor
in Student copy constructor
in Student destructor
in Student destructor
```

```
            in Student constructor
            in Student copy constructor
            in Student destructor
            in Student destructor
            1320103001    qwg     m
            1320103002    yws     m
            1320103003    wdb     m
            1320103004    hy      m
            in CDyArray destructor
            in Student destructor
            in Student destructor
            in Student destructor
            in Student destructor
```

主函数首先打开 stulist.txt 文件，读入数组元素的个数 4 并赋值给变量 num，并以 num 为参数创建数组类对象 stu，运行构造函数，输出 in CDyArray contstuctor，通过构造函数中语句 ptrArray=new Student[n]创建动态数组，共调用 4 次 Student 的构造函数，并把数组首地址赋值给 ptrArray，这里 4 次输出 in Student destructor。主函数的 for 循环，把文件中的 Student 的信息通过 CDyArray 类中的 setData 赋值给动态数组 stu 的各个元素。setData 的第 1 个形参是 Student(id,name,gen)，调用 Student 的构造函数形成实参，由于参数的传递方式是普通值传递，所以调用复制构造函数，用 Student(id,name,gen)来构造实参，显示 in Student copy constructor。CDyArray 类中的 setData 调用 Student 类中的 setData 实现对 Student 对象的赋值。由于 Student 类中的 setData 的参数传递方式为引用，形参和实参相同，所以不调用构造函数和析构函数。CDyArray 类中的 setData 运行结束，将释放创建的实参和形参，输出 2 次 in Student destructor。由于 for 循环执行 4 次，所以显示 4 次：

```
            in Student constructor
            in Student copy constructor
            in Student destructor
            in Student destructor
```

stu.print()语句将数组中的每个元素都打印出来：

```
            1320103001    qwg     m
            1320103002    yws     m
            1320103003    wdb     m
            1320103004    hy      m
```

最后，主函数运行结束，系统自动调用 CDyArray 类的析构函数，输出 in CDyArray destructor，并通过 delete []ptrArray 语句，将 4 个动态数组元素释放，运行 4 次 Student 类的析构函数，输出 4 次 in Student destructor。

3.4 类的组合成员

所谓组合关系，是指一个类中含有其他类的对象成员的情形。比如，有一个类是发动机 Moter，还有一个类是座椅 Chair。现在定义一个飞机类 Plane，其中，飞机类的成员有若干个发动机和座椅。这时可以这样定义：

```cpp
class Plane
{
    Motor *pMotor;
    Chair *pChair;
    …
public:
    Plane(int i,int j)
    {
        pMotor=new Motor[i];
        pChair=new Chair[j];
    }
    ~Plane()
    {
        delete []pMotor;
        delete []pChair;
    }

    void fly()
    {
        …
    }
    …
};
```

与组合关系相对应的还有一种关系，称为继承关系。例如，直升机类和飞机类具有继承关系，直升机继承了飞机的属性和方法，也拥有发动机和座椅，能飞行。同时，还可以增加相应的属性，如螺旋桨等。

区分组合关系和继承关系的方法是：

组合关系：是整体-部分关系(part-of)，如：座椅是飞机的一部分，发动机飞机的一部分。

继承关系：是类属关系(a-kind-of)，如：直升机是一种飞机，飞机是一种交通工具。关于继承，将在第 5 章介绍。

```cpp
/******************************************************************
    File name    :  f0304.cpp
    Description  :  类的组合
******************************************************************/
#include <iostream>
using namespace std;
class CPoint
{
public:
        CPoint(int x = 0, int y = 0)
        {
            nPosX = x;
            nPosY = y;
        }
        void print()
        {
```

```cpp
            cout<<"X="<<nPosX<<",Y="<<nPosY<<endl;
        }
    private:
        int nPosX, nPosY;
};

class Circle
{
    CPoint center;
    double radius;
    const double PI;
public:
    Circle(int x,int y,double r,double pi):
    center(x,y),radius(r),PI(pi)
    {};
    double getArea()
    {
        return PI*radius*radius;
    }
};

void main()
{
    Circle r1(1,1,1.1,3.14);
    cout<<"r1 的面积是 "<<r1.getArea()<<endl;
    Circle r2(1,1,1.1,3.141593);
    cout<<"r2 的面积是 "<<r2.getArea()<<endl;
    cout<<"二者的差是 "<<r2.getArea()-r1.getArea()<<endl;
}
```

程序的输出结果为：

```
r1 的面积是 3.7994
r2 的面积是 3.80133
二者的差是 0.00192753
```

这里要强调的是，圆和原点之间的关系是组合关系，即原点是圆的一部分。由于圆周率精度选择的不同，就会造成计算圆的面积的不同，二者的差值就是很好的体现。

再看另一个例子：

```cpp
/***************************************************************
    File name    : f0305.cpp
    Description  : 类的组合
***************************************************************/
#include <iostream>
#include <fstream>
#include <string>
using namespace std;
```

```cpp
enum Month
{
    Jan=1,Feb,Mar,Apr,May,Jun,Jul,Aug,Sep,Oct,Nov,Dec
};
class   Date
{
    int year;
    Month month;
    int day;
public:
    Date(int y=0,int m=0,int d=0):year(y),month(Month(m)),day(d)
    {}
    Date(int d,Month m,int y):year(y),month(m),day(d)
    {}
    Date(Month m,int d,int y):year(y),month(m),day(d)
    {}
    bool isLeapYear()
    {
        if( year%400==0 || (year%4==0 && year%100!=0))
            return true;
        else
            return false;
    }
    void init_date(int y,int m,int d)
    {
        year=y;
        month=Month(m);
        day=d;
    }
    void print()
    {
        cout<<year<<"-"<<month<<"-"<<day<<endl;
    }
};
//===================================

class Person
{
    string id;
    char* pName;
    char gender;
    Date birthday;
public:
    Person(string id="1320103001",char* pN="noName",
        char gender='m',int year=2000,int month=1,int day=1):
    birthday(year,month,day),id(id),gender(gender)
    {
        pName = new char[strlen(pN)+1];
```

```cpp
        if(pName)
            strcpy(pName,pN);
    }
    Person(const Person& s)
    {
        id=s.id;
        pName = new char[strlen(s.pName)+1];
        if(pName)
            strcpy(pName, s.pName);
        gender=s.gender;
    }
    void setData(string id,char *p,char gen,int year,int month,int day)
    {
        this->id=id;
        pName = new char[strlen(p)+1];
        if(pName)
            strcpy(pName, p);
        gender=gen;
        birthday.init_date(year,month,day);
    }
    ~Person()
    {
        delete[] pName;
    }
    void print()
    {
        cout<<id<<"   "<<pName<<"   "<<gender<<endl;
        birthday.print();
    }
};
void main()
{
    ifstream in("stulist1.txt");
    if (!in)
    {
        cout <<"Can't open the file!"<< endl;
        exit(0);
    }
    int num;
    in>>num;
    Person *per=new Person[num];
    string id;
    char name[20];
    char gen;
    int year,month,day;
    int i;
    for(i=0;i<num;i++)
    {
```

```
            in>>id>>name>>gen>>year>>month>>day;
            per[i].setData(id,name,gen,year,month,day);
        }
        for(i=0;i<num;i++)
        {
            per[i].print();
        }
    }
```

在该程序中，Person 类中的数据成员 birthday 不是内嵌数据类型，而是 Date 数据类型。这就是类的组合成员。类的组合成员的构造，要在该类构造函数中，通过初始化列表实现，如：

```
Person(string id="1320103001",char* pN="noName",
       char gender='m',int year=2000,int month=1,int day=1):
       birthday(year,month,day),id(id),gender(gender)
```

这里，使用数据成员的名字而不是 Date 类的名字 birthday，参数要与 Date 类构造函数的参数要求一致。

给定文件 stulist1.txt 的内容为：

```
4
1320103001   qwg m 1964 12 2
1320103002   yws m 1977  7 4
1320103003   wdb m 1983  8 8
1320103004   hy  m 1973  4 7
```

程序的输出结果为：

```
1320103001   qwg m
1964-12-2
1320103002   yws m
1977-7-4
1320103003   wdb m
1983-8-8
1320103004   hy  m
1973-4-7
```

主函数首先从 stulist1.txt 中读取人员个数 4 到变量 num，通过 Person *per=new Person[num] 语句申请堆空间，并用数组首地址初始化 per 指针。第 1 个 for 循环，利用 setData 函数对数组每个成员进行赋值，第 2 个 for 循环则将数组每个元素的信息通过 print 函数输出到屏幕。

3.5 静 态 成 员

构造函数和复制构造函数的运行都会创建对象，析构函数的运行都会销毁对象。如果希望统计当前已经生成对象的个数，该如何统计呢？

一般情况下，如果有 n 个同类的对象，那么每一个对象都分别有自己的数据成员，不同对象的数据成员各自有值，互不相干。因此，普通的数据成员无法完成上述功能。

可以使用全局变量来达到共享数据的目的，但是用全局变量的安全性得不到保证。由于在各处都可以自由地修改全局变量的值,很有可能偶然失误,全局变量的值就会修改,导致程序的结果出错误。

```cpp
/******************************************************************
    File name    : f0306.cpp
    Description  : 利用全局变量统计生成对象个数
******************************************************************/
#include <iostream>
using namespace std;
int countP=0;          //全局变量,用于统计创建的 Point 对象个数
class Point                                //Point 类声明
{
public:                                    //外部接口
    Point(int xx=0, int yy=0)              //构造函数
    {
        X=xx;
        Y=yy;
        countP++;
        cout<<countP<<endl;
    }
    Point(Point &p);                       //复制构造函数
    ~Point()
    {
        countP--;
        cout<<countP<<endl;
    }
    int GetX()
    {
        return X;
    }
    int GetY()
    {
        return Y;
    }
private:                                   //私有数据成员
    int X,Y;
};
Point::Point(Point &p)
{
    X=p.X;
    Y=p.Y;
    countP++;
    cout<<countP<<endl;
}

void main()                                //主函数实现
{
    cout<<countP<<endl;
```

```cpp
    Point A(4,5);                              //声明对象A
    cout<<"Point A,"<<A.GetX()<<","<<A.GetY();
    cout<<" Object id="<<++countP<<endl;
    Point B(A);                                //声明对象B
    cout<<"Point B,"<<B.GetX()<<","<<B.GetY();
    cout<<" Object id="<<countP<<endl;
}
```

程序的输出结果为:

```
0
1
Point A,4,5 Object id=2
3
Point B,4,5 Object id=3
2
1
```

这里，主函数 main 中的++countP 导致统计的结果发生错误，因此在实际工作中很少使用全局变量。由于 countP 是全局变量，在程序的任何函数中都有可能使用它，这就很容易发生错误。如果想在同类的多个对象之间实现数据共享，但不用全局对象，可以用类的静态数据成员。

3.5.1 静态数据成员

静态数据成员是一种特殊的数据成员。它以关键字 static 开头。例如：

```cpp
/*******************************************************************
    File name   : f0307.cpp
    Description : 静态成员变量的使用
*******************************************************************/
#include <iostream>
using namespace std;
class Point                                //Point 类声明
{
public:                                    //外部接口
    Point(int xx=0, int yy=0)
    {
        X=xx;
        Y=yy;
        countP++;
        cout<<countP<<endl;
    }                                      //构造函数
    Point(Point &p);                       //复制构造函数
    ~Point()
    {
        countP--;
        cout<<countP<<endl;
    }
    int GetX()
```

```
        {
            return X;
        }
        int GetY()
        {
            return Y;
        }
        static int countP;              //静态数据成员引用性说明
    private:                            //私有数据成员
        int X,Y;
};
Point::Point(Point &p)
{
    X=p.X;
    Y=p.Y;
    countP++;
    cout<<countP<<endl;
}

int Point::countP=0;                    //静态数据成员定义性说明

void main()                             //主函数实现
{
    cout<<Point::countP<<endl;
    Point A(4,5);                       //声明对象A
    cout<<"Point A,"<<A.GetX()<<","<<A.GetY();
    cout<<" Object id="<<Point::countP<<endl;
    Point B(A);                         //声明对象B
    cout<<"Point B,"<<B.GetX()<<","<<B.GetY();
    cout<<" Object id="<<Point::countP<<endl;
}
```

程序的输出结果为：

```
0
1
Point A,4,5 Object id=1
2
Point B,4,5 Object id=2
1
0
```

　　静态数据成员为各对象所共有，而不只属于某个对象的成员，所有对象都可以引用它。甚至在没有对象出现时，该静态数据成员也是存在的。静态的数据成员在内存中只占一份空间。每个对象都可以引用这个静态数据成员。静态数据成员的值对所有对象都是一样的。如果改变它的值，则在各对象中这个数据成员的值都同时改变。当然，把静态变量 countP 定义成 public 还是存在危险的，这样也造成该成员变量可以在程序任意位置使用它，也容易产生错误。希望将其定义成 private，利用函数使用它。

将 countP 定义成 private，并定义 public 成员函数 showP：

```
void showP()
{
    Cout<<countP<<endl;
}
```

这样就可以将主函数改成：

```
void main()                         //主函数实现
{
//  cout<<Point::countP<<endl;
    Point A(4,5);                   //声明对象 A
    cout<<"Point A,"<<A.GetX()<<","<<A.GetY();
    cout<<" Object id="<<A.showP()<<endl;
    Point B(A);                     //声明对象 B
    cout<<"Point B,"<<B.GetX()<<","<<B.GetY();
    cout<<" Object id="<<B.showP()<<endl;
    cout<<" Object id="<<A.showP()<<endl;
}
```

程序的输出结果为：

```
1
Point A,4,5 Object id=1
2
Point B,4,5 Object id=2
Object id=2
1
0
```

程序首先创建对象 A，构造函数输出 1，cout<<"Point A,"<<A.GetX()<<","<<A.GetY();输出 Point A,4,5，由于 showP 是成员函数，可以利用语句 A.showP()来显示 Object id=1。Point B(A)语句调用复制构造函数创建 B，复制构造函数输出 2，cout<<"Point B,"<<B.GetX()<<",","<<B.GetY();输出 Point A,4,5，后面的 B.showP()和 A.showP()都是输出的 2。可以看出，利用成员变量来输出静态成员，很好地保证了结果的正确性，减小了出错的可能。但是，静态成员变量是在对象创建前就存在的，普通成员函数的调用依附于创建的对象，所以无法显示静态成员变量 countP 为 0 的情况。普通成员函数无法实现，将在 3.5.2 节引入静态成员函数来解决上述问题。

类的数据成员有普通数据成员和静态数据成员两类。在实际应用中该如何区分两者呢？有些成员变量是针对每个对象个体描述的，这是普通的成员变量。例如，想建立一个熊猫类，每个熊猫个体，都有供人区分的名字、性别、出生日期(如果是年龄，一定要有对象创建时间，否则年龄将随着时间的改变而失效)等。同时，有一些信息则是属于整个熊猫类的，如熊猫的生物学分类，学名：大熊猫，拉丁学名：Ailuropoda melanoleuca，英文学名：Giant Panda。这样的成员变量应该定义为静态数据成员。静态数据成员不是类对象的属性，而是整个类的属性。

用户可以这样定义一个熊猫类：

```
/************************************************************
File name   : f0308.cpp
```

```
Description :  静态成员变量的使用
*********************************************************************/
#include <iostream>
#include <string>
using namespace std;

class Panda
{
    static string ChName;
    static string EnName;
    static string LtName;
    string name;
    char sex;
    string place;
    int yearofBirth;
public:
    Panda(string nm="noName",char sx='m',
        string place="四川",int year=2015):
    name(nm),sex(sx),place(place),yearofBirth(year)
    {}
    void print()
    {
        cout<<"名字："<< name << endl
            <<"性别："<<( (sex=='m') ? "雄性":"雌性")<<endl
            <<"地点："<< place<<endl
            <<"出生年份："<<yearofBirth<<endl;
    }
    static void intro()
    {
        cout<<"熊猫的"<<endl
            <<"中文名称："<<ChName<<endl
            <<"英文名称："<<EnName<<endl
            <<"拉丁名称："<<LtName<<endl;
    }
};
string Panda::ChName="大熊猫";
string Panda::EnName="Giant Panda";
string Panda::LtName="Ailuropoda melanoleuca";
void main()
{
    Panda panpan("巴斯",'m',"洛杉矶",1980);
    Panda xiwang("希望",'f',"四川",2005);
    panpan.intro();
    panpan.print();
    xiwang.intro();
    xiwang.print();
}
```

程序的输出结果为：

```
熊猫的
中文名称：大熊猫
英文名称：Giant Panda
拉丁名称：Ailuropoda melanoleuca
名字：巴斯
性别：雄性
地点：洛杉矶
出生年份：1980
熊猫的
中文名称：大熊猫
英文名称：Giant Panda
拉丁名称：Ailuropoda melanoleuca
名字：希望
性别：雌性
地点：四川
出生年份：2005
```

关于静态数据成员的几点说明：

(1)如果只声明了类而未定义对象，则类的一般数据成员是不占内存空间的，只有在定义对象时，才为对象的数据成员分配空间。但是静态数据成员不属于某一个对象，在为对象所分配的空间中不包括静态数据成员所占的空间。静态数据成员是在所有对象之外单独开辟空间。只要在类中定义了静态数据成员，即使不定义对象，也为静态数据成员分配空间，它可以被引用。静态成员变量存于全局数据区。

(2)如果一个函数定义了静态变量，在函数结束时该静态变量并不释放，仍然存在并保留其值。静态数据成员也类似，它不随对象的建立而分配空间，也不随对象的撤销而释放空间。静态数据成员是在程序编译时分配空间的，到程序结束时才释放空间，如果未对静态数据成员赋初值，则编译系统会自动赋予初值 0 或者空。

(3)静态数据成员可以初始化，但只能在类体外进行初始化。例如：

```
string Panda::ChName="大熊猫";
string Panda::EnName="Giant Panda";
string Panda::LtName="Ailuropoda melanoleuca";
```

用于表示 Panda 类的共同属性。其一般形式为：

```
数据类型类名::静态数据成员名=初值;
```

不能在初始化语句中加 static。**注意，不能用构造函数的参数对静态数据成员初始化。**

(4)静态数据成员既可以通过对象名引用，也可以通过类名来引用。观察下面的程序：将 static string ChName;的访问权限定义成 public，运行下面主函数：

```
void main()
{
    cout<<Panda::ChName<<endl;
    Panda basi("巴斯",'m',"洛杉矶",1980);
    Panda xiwang("希望",'f',"四川",2005);
    cout<<basi.ChName<<endl;
    cout<<xiwang.ChName<<endl;
```

```
        cout<<Panda::ChName<<endl;
}
```

程序的输出结果为:

```
大熊猫
大熊猫
大熊猫
大熊猫
```

上面 4 个输出语句的输出结果相同,这就验证了所有对象的静态数据成员实际上是同一个数据成员。可以看到在类外通过对象名引用公用的静态数据成员,也可以通过类名引用静态数据成员。即使没有定义类对象,也可以通过类名引用静态数据成员。这说明静态数据成员并不是属于对象的,而是属于类的。类的对象也可以引用它,但这并不代表该静态成员属于这个对象。如果静态数据成员被定义为私有的,则不能在类外直接引用,必须通过公用的成员函数引用。

(5)有了静态数据成员,各对象之间的数据有了沟通的渠道,实现整个类的数据共享。注意,公用静态数据成员与全局变量的不同,静态数据成员的作用域只限于定义该类的作用域内。在此作用域内,可以通过类名和域运算符":::"引用静态数据成员,而不论类对象是否存在。

3.5.2 静态成员函数

在程序 f0308.cpp 中,用函数

```
        void intro();
```

来显示静态变量信息。虽然可以实现对静态变量的显示,但是必须用对象调用该函数。这样,就不能实现像:

```
        cout<<Panda::ChName<<endl;
```

那样,在没有对象形成之前就可以对静态信息进行显示。事实上,这些信息是与对象无关的。假如将 ChName 重新定义为 private 访问权限,又想通过访问权限为 public 的函数对其进行处理。此时,普通的成员函数无法实现。

C++提供了**静态成员函数**,在类中声明函数的前面加 static 就成了静态成员函数。如

```
        static void intro();
```

和静态数据成员一样,静态成员函数是类的一部分,而不是对象的一部分。如果要在类外调用公用的静态成员函数,要用类名和域运算符"::"。例如:

```
        Panda:: intro();
```

静态成员函数也允许通过对象名调用静态成员函数。例如:

```
        Panda basi("巴斯",'m',"洛杉矶",1980);
        basi.intro();
```

但这并不意味着此函数是属于对象 basi 的,而只是用 basi 的类型而已。

静态成员函数的作用是为了能处理静态数据成员。在 C++程序中,静态成员函数主要用来访问静态数据成员,而不访问非静态成员。静态成员函数可以直接引用本类中的静态数据成员,因为静态成员同样是属于类的,可以直接引用。

当调用一个对象的非静态成员函数时，系统会把该对象的起始地址赋给成员函数的 this 指针。而静态成员函数并不属于某一对象，它与任何对象都无关，因此静态成员函数没有 this 指针。既然它没有指向某一对象，就无法对一个对象中的非静态成员进行默认访问，即在引用数据成员时不指定对象名。

静态成员函数与非静态成员函数的根本区别是：**非静态成员函数有 this 指针，而静态成员函数没有 this 指针**。由此决定了静态成员函数不能访问本类中的非静态成员。静态成员函数：

```
static void intro()
{
        cout<<"熊猫的"<<endl
            <<"中文名称："<<ChName<<endl
            <<"英文名称："<<EnName<<endl
            <<"拉丁名称："<<LtName<<endl;
}
```

不能直接输出类中的非静态成员：

```
string name;
char sex;
string place;
int yearofBirth;
```

静态成员函数没有 this 指针，因此无法确定是哪个对象的这些成员变量。这也是为什么定义 print 函数来输出这些数据成员的原因。

作为 C++程序员，最好养成这样的习惯：只用静态成员函数引用静态数据成员，而不引用非静态数据成员。这样思路清晰，逻辑清楚，不易出错。

通过下面的例子可以具体了解有关引用非静态成员的具体方法。在 f0307.cpp 中加入：

```
static void showCount()
{
    cout<<"Count="<<countP<<endl;
}
```

主函数改成：

```
void main()
{
    Point::showCount();            //声明一个 int 型指针，指向类的静态成员
    Point A(4,5);                  //声明对象 A
    cout<<"Point A,"<<A.GetX()<<","<<A.GetY()<<endl;
    Point::showCount();            //直接通过指针访问静态数据成员
    Point B(A);                    //声明对象 B
    cout<<"Point B,"<<B.GetX()<<","<<B.GetY()<<endl;
    Point::showCount();            //直接通过指针访问静态数据成员
    A.showCount();
    B.showCount();
}
```

程序的输出结果为：

```
Count=0
```

```
1
Point A,4,5
Count=1
2
Point B,4,5
Count=2
Count=2
Count=2
1
0
```

可以看出，利用静态成员函数处理静态成员变量是最佳的方法，静态成员函数可以随时访问静态成员变量，而与是否创建类对象无关。

3.6 友　元

类的数据隐藏特性，通过将成员变量设定为 private 或者 protected 访问权限，将其与类外用户的使用隔离开来。如果想使用这些成员变量，则需要通过被定义成 public 访问权限，被称为接口函数的成员函数来操作成员变量。数据隐藏为高质量的程序设计带来好处。

同时也应看到，原本可以直接访问的变量，现在需要通过接口函数访问，而函数的运行需要参数传递、现场保护、函数运行、现场恢复等一系列的工作，需要消耗许多时间和空间。这对那些对时空效率非常重视的任务，如实时数据处理和科学计算来说，将是难以接受的。

为了提高效率，C++提供了特殊的机制，可以让普通函数直接访问类的私有成员和保护成员，需要把该函数定义成该类的友元函数。

友元函数是说明在类体内的一般函数，也可以是另一个类的成员函数。友元函数并不是这个类中的成员函数，但是它却可以直接访问该类中所有成员，包括 private 和 protected 的成员。

3.6.1 普通函数作为友元函数

友元函数说明格式如下：

```
friend <类型> <函数名>(<参数列表>)
```

此例是用于做一个矩阵与列向量的乘法，矩阵类和向量类分别为 Matrix 和 Vector。下面的程序中 multiply 函数是将乘法作为普通函数，multiply1 则是友元函数。

```
/******************************************************************
File name    :  f0309.cpp
Description  :  普通函数作为友元函数
******************************************************************/
#include<iostream>
#include<fstream>
#include <ctime>
using namespace std;
//-------------------------------------
class Vector;
```

```cpp
class Matrix;
class Vector
{
    int* v;                              //指向一个数组,表示向量
    int sz;
public:
    friend Vector multiply1(Matrix& m, Vector& v);
    void remove()
    {
    delete[] v;
    }
    int size()
{
return sz;
}
    void set(int);
    void display();
    int& operator[](int);
};
//---------------------------------
void Vector::set(int s)
{
    sz = s;
    if(s<=0)
    {
        cerr<<"bad Vector size.\n";
        exit(1);
    }
    v = new int[s];
}//-----------------------------------
int& Vector::operator[](int i)           //引用返回的目的是返回值可以作左值
{
    if(i<0 || i>=sz)
    {
        cerr <<"Vector index out of range.\n";
        exit(1);
    }
    return v[i];
}//-----------------------------------
void Vector::display()
{
    for(int i=0; i<sz; ++i)
        cout<<v[i]<<" ";
        cout<<"\n";
}//-----------------------------------
class Matrix
{
    int* m;
```

```cpp
    int szl, szr;
    friend Vector multiply1(Matrix& m, Vector& v);
public:
    void set(int, int);
    void remove()
    {
        delete[] m;
    }
    inline int sizeL();
    inline int sizeR();
    inline int& elem(int, int);
};//-----------------------------------
int Matrix::sizeL()
{
    return szl;
}
int Matrix::sizeR()
{
    return szr;
}
void Matrix::set(int i, int j)
{
    szl = i;
    szr = j;
    if(i<=0 || j<=0)
    {
        cerr <<"bad Matrix size.\n";
        exit(1);
    }
    m = new int[i*j];
}//-----------------------------------
int& Matrix::elem(int i, int j)        //引用返回的目的是返回值可以作左值
{
    if(i<0||szl<=i||j<0||szr<=j)
    {
        cerr <<"Matrix index out of range.\n";
        exit(1);
    }
    return m[i*szr+j];
}//-----------------------------------
Vector multiply(Matrix& m, Vector& v) //矩阵乘向量
{
    if(m.sizeR()!=v.size())
    {
        cerr <<"bad multiply Matrix with Vector.\n";
        exit(1);
    }
    Vector r;                          //创建一个存放结果的空向量
```

```cpp
        r.set(m.sizeL());
        for(int i=0; i<m.sizeL(); i++)
        {
            r[i] = 0;
            for(int j=0; j<m.sizeR(); j++)
                r[i] += m.elem(i,j) * v[j];
        }
        return r;
}//-----------------------------------
Vector multiply1(Matrix& m, Vector& v)  //矩阵乘向量
{
        if(m.szr!=v.sz)
        {
            cerr <<"bad multiply Matrix with Vector.\n";
            exit(1);
        }
        Vector r;                                          //创建一个存放结果的空向量
        r.set(m.szl);
        for(int i=0; i<m.szl; i++)
        {
            r[i] = 0;
            for(int j=0; j<m.szr; j++)
                r[i] += m.m[i*m.szr+j] * v[j];
        }
        return r;
}//-----------------------------------
void main()
{
        ifstream in("in.txt");
    if (!in)
        {
            cout <<"Can't open the file!"<< endl;
            exit(0);
        }
        int x,y;
        in>>x>>y;
        Matrix ma;
        ma.set(x,y);
        for(int i=0; i<x; ++i)
            for(int j=0; j<y; ++j)
                in>>ma.elem(i,j);
        in>>x;
        Vector ve;
        ve.set(x);
        for(i=0; i<x; ++i)
            in>>ve[i];
        Vector vx;
        double t_start = clock();
```

```
        for(i=0;i<10000000;i++)
            vx = multiply(ma,ve);
        cout << clock() - t_start << endl;
        vx.display();
        t_start = clock();
        for(i=0;i<10000000;i++)
            vx = multiply1(ma,ve);
        cout << clock() - t_start << endl;
        vx.display();
        ma.remove();
        ve.remove();
        vx.remove();
    }
    //====================================
```

当 in.txt 的内容为：

```
4 3
1 2 3
0 1 2
1 1 3
1 2 1
3
2 1 0
```

程序的运行结果为：

```
16497
4 1 3 4
9420
4 1 3 4
```

程序运行时，先读一个矩阵，再读一个向量。读矩阵时，先读行和列数，根据行列数，通过 set 函数设置矩阵空间的大小(行×列)。用一个两重循环，调用 elem 函数读入矩阵元素。读向量时，先读列数，根据列数，通过 set 函数设置向量空间的大小(列)。用一个循环读入向量元素。

可以看到，multiply 函数中，使用了矩阵和向量中的成员函数：

```
        m.sizeR();m.sizeL();m.elem(i,j);v.size();
```

来获取相应的私有数据。

由于计算机运行速度非常快，为了能够展示友元函数的良好性能，将乘法运算循环运行了 10000000。可以发现，利用友元函数之后，这些数据成员可以直接访问，运行时间由 16497 减少到 9420，大大提高了运行效率。当然，在不同机器上运行，用时可能不同，但是二者的比例大致相当。

3.6.2 成员函数作为友元函数

友元函数既可以是普通函数，也可以是某一类的成员函数。如果该友元函数是某一类的成员函数，则需要在声明时，在函数名前加上类名，以明确所属的类名。说明格式如下：

```
friend <类型> <类名::函数名>(<参数列表>)

/*********************************************************************
File name   :  f0310.cpp
Description :  成员函数作为友元函数
*********************************************************************/
#include<iostream>
#include<fstream>
#include <ctime>
using namespace std;
//-------------------------------------
class Vector;
class Matrix;

class Matrix
{
    int* m;
    int szl, szr;
    friend Vector multiply1(Matrix& m, Vector& v);
public:
    Vector multiply11(Vector& v);
    void set(int, int);
    void remove()
    {
        delete[] m;
    }
    inline int sizeL();
    inline int sizeR();
    inline int& elem(int, int);
};//-----------------------------------
int Matrix::sizeL()
{
    return szl;
}
int Matrix::sizeR()
{
    return szr;
}
void Matrix::set(int i, int j)
{
    szl = i;
    szr = j;
    if(i<=0 || j<=0)
    {
        cerr<<"bad Matrix size.\n";
        exit(1);
    }
    m = new int[i*j];
```

```cpp
}//-----------------------------------
int& Matrix::elem(int i, int j)         //引用返回的目的是返回值可以作左值
{
    if(i<0||szl<=i||j<0||szr<=j)
    {
        cerr<<"Matrix index out of range.\n";
        exit(1);
    }
    return m[i*szr+j];
}//-----------------------------------

class Vector
{
    int* v;                             //指向一个数组,表示向量
    int sz;
    friend Vector multiply1(Matrix& m, Vector& v);
public:
    friend Vector Matrix::multiply11(Vector& v);
    void remove()
    {
        delete[] v;
    }
    int size()
    {
        return sz;
    }
    void set(int);
    void display();
    int& operator[](int);
};//-----------------------------------
void Vector::set(int s)
{
    sz = s;
    if(s<=0)
    {
        cerr<<"bad Vector size.\n";
        exit(1);
    }
    v = new int[s];
}//-----------------------------------
int& Vector::operator[](int i)          //引用返回的目的是返回值可以作左值
{
    if(i<0 || i>=sz)
    {
        cerr<<"Vector index out of range.\n";
        exit(1);
    }
    return v[i];
```

```cpp
}//----------------------------------
void Vector::display()
{
    for(int i=0; i<sz; ++i)
        cout<<v[i]<<" ";
    cout<<"\n";
}//----------------------------------
Vector multiply(Matrix& m, Vector& v)    //矩阵乘向量
{
    if(m.sizeR()!=v.size())
    {
        cerr<<"bad multiply Matrix with Vector.\n";
        exit(1);
    }
    Vector r;
    r.set(m.sizeL());                       //创建一个存放结果的空向量
    for(int i=0; i<m.sizeL(); i++)
    {
        r[i] = 0;
        for(int j=0; j<m.sizeR(); j++)
            r[i] += m.elem(i,j) * v[j];
    }
    return r;
}//----------------------------------
Vector multiply1(Matrix& m, Vector& v)    //矩阵乘向量
{
    if(m.szr!=v.sz)
    {
        cerr<<"bad multiply Matrix with Vector.\n";
        exit(1);
    }
    Vector r;
    r.set(m.szL());                         //创建一个存放结果的空向量
    for(int i=0; i<m.szl; i++)
    {
        r[i] = 0;
        for(int j=0; j<m.szr; j++)
            r[i] += m.m[i*m.szr+j] * v[j];
    }
    return r;
}//----------------------------------
Vector Matrix::multiply11(Vector& v)    //矩阵乘向量
{
    if(szr!=v.sz)
    {
        cerr<<"bad multiply Matrix with Vector.\n";
        exit(1);
    }
```

```cpp
        Vector r;
        r.set(szl);                              //创建一个存放结果的空向量
        for(int i=0; i<szl; i++)
        {
            r[i] = 0;
            for(int j=0; j<szr; j++)
                r[i] += m[i*szr+j] * v[j];
        }
        return r;
}//----------------------------------
void main()
{
    Ifstream in("in.txt");
        if (!in)
        {
            cout <<"Can't open the file!"<< endl;
            exit(0);
        }
    int x,y;
    in>>x>>y;
    Matrix ma;
    ma.set(x,y);
    for(int i=0; i<x; ++i)
        for(int j=0; j<y; ++j)
            in>>ma.elem(i,j);
    in>>x;
    Vector ve;
    ve.set(x);
    for(i=0; i<x; ++i)
        in>>ve[i];
    Vector vx;
    double t_start = clock();
    for(i=0;i<10000000;i++)
        vx = multiply(ma,ve);
    cout<< clock() - t_start << endl;
    vx.display();
    t_start = clock();
    for(i=0;i<10000000;i++)
        vx = ma.multiply11(ve);
    cout<< clock() - t_start << endl;
    vx.display();
    ma.remove();
    ve.remove();
    vx.remove();
}//==================================
```

在该程序中，在 **Vector** 类中声明：

```cpp
    friend Vector Matrix::multiply11(Vector& v);
```

说明 multiply11 是 Matrix 类的成员函数，同时又是 Vector 的友元函数。于是，该函数也可以直接使用 Vector 中的每个成员，而不受访问权限的制约。

当然，该乘法也可以定义成 Vector 的成员函数。此时，需要在 Matrix 中声明

```
friend Vector Vector::multiply111(Matrix& M);
```

说明 multiply111 是 Vector 类的成员函数，同时又是 Matrix 的友元函数。于是，该函数也可以直接使用 Matrix 中的每个成员，而不受访问权限的制约。

```
Vector Vector::multiply1111(Matrix& m)    //矩阵乘向量
{
    if(m.szr!=sz)
    {
        cerr<<"bad multiply Matrix with Vector.\n";
        exit(1);
    }
    Vector r;
    r.set(m.szl);                          //创建一个存放结果的空向量
    for(int i=0; i<m.szl; i++)
    {
    r[i] = 0;
    for(int j=0; j<m.szr; j++)
        r[i] += m.m[i*m.szr+j] * v[j];
    }
    return r;
}
```

这里，向量作为当前对象，矩阵则成为参数 ve.multiply1111(ma)。虽然可以实现乘法的功能，但是向量在前，矩阵作为参数，不如使用 ma.multiply11(ve) 方式，矩阵作为当前对象，向量作为参数那样更符合人们的认知习惯。

3.6.3 友元类

除了普通函数作为友元函数，一个类的成员函数作为另一个类的友元函数之外，某个类也可以定义成另一个类的友元，这个类称为友元类。友元类的每个成员函数都可以访问另一个类中的保护或者私有成员。

对于矩阵和向量乘法的例子，也可以使用友元类来实现。

```
/*****************************************************************
    File name   :  f0311.cpp
    Description :  友元类
*****************************************************************/
#include<iostream>
#include<fstream>
#include <ctime>
using namespace std;
//-----------------------------------
class Vector;
class Matrix;
```

```cpp
class Matrix
{
    int* m;
    int szl, szr;
public:
    Vector multiply11111(Vector& v);
    void set(int, int);
    void remove()
    {
        delete[] m;
    }
    inline int sizeL();
    inline int sizeR();
    inline  int& elem(int, int);
};//----------------------------------
int Matrix::sizeL()
{
    return szl;
}
int Matrix::sizeR()
{
    return szr;
}
void Matrix::set(int i, int j)
{
    szl = i;
    szr = j;
    if(i<=0 || j<=0)
    {
        cerr<<"bad Matrix size.\n";
        exit(1);
    }
    m = new int[i*j];
}//--------------------------------------
int& Matrix::elem(int i, int j)         //引用返回的目的是返回值可以作左值
{
    if(i<0||szl<=i||j<0||szr<=j)
    {
        cerr<<"Matrix index out of range.\n";
        exit(1);
    }
    return m[i*szr+j];
}//--------------------------------------

class Vector
{
    int* v;                                           //指向一个数组,表示向量
```

```cpp
        int sz;
public:
    friend Matrix;
    void remove()
    {
        delete[] v;
    }
    int size()
    {
        return sz;
    }
    void set(int);
    void display();
    int& operator[](int);
};//---------------------------------
void Vector::set(int s)
{
    sz = s;
    if(s<=0)
    {
        cerr<<"bad Vector size.\n";
        exit(1);
    }
    v = new int[s];
}//---------------------------------
int& Vector::operator[](int i)        //引用返回的目的是返回值可以作左值
{
    if(i<0 || i>=sz)
    {
        cerr<<"Vector index out of range.\n";
        exit(1);
    }
    return v[i];
}//---------------------------------
void Vector::display()
{
    for(int i=0; i<sz; ++i)
        cout<<v[i]<<" ";
    cout<<"\n";
}//---------------------------------
Vector multiply(Matrix& m, Vector& v)    //矩阵乘向量
{
    if(m.sizeR()!=v.size())
    {
        cerr<<"bad multiply Matrix with Vector.\n";
        exit(1);
    }
    Vector r;
```

```cpp
        r.set(m.sizeL());                      //创建一个存放结果的空向量
        for(int i=0; i<m.sizeL(); i++)
        {
            r[i] = 0;
            for(int j=0; j<m.sizeR(); j++)
                r[i] += m.elem(i,j) * v[j];
        }
        return r;
}//-----------------------------------
Vector Matrix::multiply11111(Vector& v)        //矩阵乘向量
{
    if(szr!=v.sz)
    {
        cerr<<"bad multiply Matrix with Vector.\n";
        exit(1);
    }
    Vector r;
    r.set(szl);                                //创建一个存放结果的空向量
    for(int i=0; i<szl; i++)
    {
        r[i] = 0;
        for(int j=0; j<szr; j++)
            r[i] += m[i*szr+j] * v[j];
    }
    return r;
}//-----------------------------------


void main()
{
    ifstream in("in.txt");
    if (!in)
    {
        cout <<"Can't open the file!"<< endl;
        exit(0);
    }
    int x,y;
    in>>x>>y;
    Matrix ma;
    ma.set(x,y);
    for(int i=0; i<x; ++i)
        for(int j=0; j<y; ++j)
            in>>ma.elem(i,j);
    in>>x;
    Vector ve;
    ve.set(x);
    for(i=0; i<x; ++i)
        in>>ve[i];
```

```
    Vector vx;
    double t_start = clock();
    for(i=0;i<10000000;i++)
        vx = multiply(ma,ve);
    cout<< clock() - t_start << endl;
    vx.display();
    t_start = clock();
    for(i=0;i<10000000;i++)
        vx = ma.multiply11111(ve);
    cout<< clock() - t_start << endl;
    vx.display();
    ma.remove();
    ve.remove();
    vx.remove();
}
//===========================
```

3.7 综合应用例题

链表是一种重要的数据结构，在程序设计中广泛使用。这里通过建立链表类来实现链表的功能。

```
/******************************************************************
    File name   : listP.h
    Description : 链表类的头文件
******************************************************************/
typedef int listItemType;
struct listNode;
typedef listNode* ptrType;
class listClass
{
private:
    int Size;
    ptrType Head;
    ptrType PtrTo(int Position)const;
public:
    listClass();
    listClass(const listClass& L);
    ~listClass();
    bool ListIsEmpty()const;
    int ListLength()const;
    void ListInsert(int NewPosition,listItemType NewItem,bool& Success);
    void ListDelete(int Position,bool& Success);
    void ListRetrieve(int Position,listItemType& DataItem,bool& Success)const;
};
/******************************************************************
    File name  :  listp.cpp
```

```
      Description :  链表类的实现文件
******************************************************************/

#include "ListP.h"
#include <stddef.h>
#include <assert.h>
struct listNode
{
    listItemType Item;
    ptrType Next;
};

listClass::listClass():Size(0),Head(NULL)
{}

listClass::listClass(const listClass& L):Size(L.Size)
{
    if(L.Head==NULL)
        Head=NULL;
    else
    {
        Head=new listNode;
        assert(Head!=NULL);
        Head->Item=L.Head->Item;
        ptrType NewPtr=Head;
        for(ptrType OrigPtr=L.Head->Next;OrigPtr!=NULL;OrigPtr=OrigPtr->Next)
        {
            NewPtr->Next=new listNode;
            assert(NewPtr->Next!=NULL);
            NewPtr=NewPtr->Next;
            NewPtr->Item=OrigPtr->Item;
        }
        NewPtr->Next=NULL;
    }
}

listClass::~listClass()
{
    bool Success;
    while(!ListIsEmpty())
        ListDelete(1,Success);
}
bool listClass::ListIsEmpty()const
{
    return bool(Size==0);
}
int listClass::ListLength()const
{
```

```cpp
        return Size;
}
ptrType listClass::PtrTo(int Position)const
{
    if((Position<1)||(Position>ListLength()))
        return NULL;
    else
    {
        ptrType Cur=Head;
        for(int skip=1;skip<Position;skip++)
            Cur=Cur->Next;
        return Cur;
    }
}
void listClass::ListInsert(int NewPosition,listItemType NewItem,bool& Success)
{
    int NewLength=ListLength()+1;
    Success=bool((NewPosition>=1)&&(NewPosition<=NewLength));
    if(Success)
    {
        ptrType NewPtr=new listNode;
        Success=bool(NewPtr!=NULL);
        if(Success)
        {
            Size=NewLength;
            NewPtr->Item=NewItem;
            if(NewPosition==1)
            {
                NewPtr->Next=Head;
                Head=NewPtr;
            }
            else
            {
                ptrType Prev=PtrTo(NewPosition-1);
                NewPtr->Next=Prev->Next;
                Prev->Next=NewPtr;
            }
        }
    }
}
void listClass::ListDelete(int Position,bool& Success)
{
    ptrType Cur;
    Success=bool((Position>=1)&&(Position<=Size));
    if(Success)
    {
        Size--;
        if(Position==1)
```

```cpp
            {
                Cur=Head;
                Head=Head->Next;
            }
            else
            {
                ptrType Prev=PtrTo(Position-1);
                Cur=Prev->Next;
                Prev=Cur->Next;
            }
            delete Cur;
            Cur=NULL;
    }
}
void listClass::ListRetrieve(int Position,listItemType& DataItem,bool& Success)const
{
    Success=bool((Position>=1)&&(Position<=Size));
    if(Success)
    {
        ptrType Cur=PtrTo(Position);
        DataItem=Cur->Item;
    }
}

/******************************************************************
    File name    : f0312.cpp
    Description  : 链表类应用的主函数
******************************************************************/

#include "ListP.h"
#include <iostream.h>
void main()
{
    listClass L;
    listItemType DataItem;
    bool Success;
    L.ListInsert(1,10,Success);
    L.ListInsert(2,20,Success);
    L.ListInsert(3,30,Success);
    for(int i=1;i<=L.ListLength();i++)
    {
        L.ListRetrieve(i,DataItem,Success);
        cout<<DataItem<<endl;
    }
    L.ListDelete(1,Success);
    for(i=1;i<=L.ListLength();i++)
    {
        L.ListRetrieve(i,DataItem,Success);
```

```
            cout<<DataItem<<endl;
        }
    }
```

练　习

1．试建立一个学生类，除具有姓名、学号、出生年月等信息，还要有学生的修课记录，要求按照学期记录，根据学期选课情况动态申请空间。修课记录包括：学生的课程名、教师、成绩等信息。

2．在练习1的基础上，建立班级类，按照班级对学生进行管理，所有信息存入文件，要求对班级学生以及对学生的成绩进行检索，并统计学生的平均成绩，按照降序排列。

3．本章矩阵与列向量的乘法的例题，都是在很小规模的矩阵和列向量上进行的。试首先随机建立一个高维的矩阵和向量，然后重复友元函数的程序。

4．书中链表类管理的是整型数据。加入利用链表管理学生类的数据，需要如何修改程序？请实现之。

第 4 章 运算符重载

运算符重载是面向对象程序设计中最吸引人的特征之一,其目的是将人们日常所用的运算符,以自然的方式扩展到 C++语言。将复杂难理解的程序变得更加直观、更加符合人的思维。运算符重载增强了 C++的可扩展性。本章主要讲解运算符重载的基本概念、规则、方法特点及应用。

- 知识要点

运算符重载原理;
运算符重载的方法、规则;
单目运算符重载、双目运算符重载、类型转换。

- 探索思考

运算符重载给程序设计带来什么好处?有什么优点?
深复制和浅复制的区别?什么时候需要深复制?

- 预备知识

类的定义、构造函数、析构函数、深复制和浅复制。

4.1 运算符重载的定义

C++中预定义运算符的操作对象只能是基本数据类型,用户自定义类型如果需要相应运算操作,只能定义相关成员函数。例如,有复数类 CComplex 定义如下:

```cpp
/******************************************************************
    File name   : f0401.cpp
    Description : 复数类定义
******************************************************************/
class CComplex
{
private:
    double m_real, m_image;
public:
    CComplex(double r, double i)
    {
        m_real = r;
        m_image = i;
    }
    CComplex add(const CComplex& c)
    {
        CComplex temp(*this);
        temp.m_real += c.m_real;
```

```
            temp.m_image += c.m_image;
            return temp;
        }
};
//================================================================
```

现在要计算两个复数的和，并将结果保存在另外一个复数对象中，则只能用如下代码实现：

```
CComplex c1(1,2), c2(3,4);
CComplex c3 = c1.add(c2);
```

如果能为 CComplex 类提供如下所示的加法运算：

```
CComplex c3=c1+c2;
```

显然上面形式与成员函数调用相比，更加直观，也更加容易理解，这就是运算符重载。C++语言提供了运算符重载功能，使得用户可以赋予运算符新的功能，使运算符能够用于特定类型执行特定的操作。这种重新定义运算符，使其能够用于特定的类对象完成特定操作的技术称为运算符重载。运算符重载的实质是函数重载，它将运算符重载为具有特定名称和参数的函数，其运行过程仍然是函数调用过程。运算符重载提供了 C++ 的可扩展性，也是 C++ 最吸引人的特性之一。

运算符重载可以方便编程和程序直观上的理解，但运算符重载并不是必需的，主要是为了在编程中进行人性化描述，达到更方便地理解程序的目的。

4.2 运算符重载的方法

C++中运算符既可以重载为类的非静态成员函数，也可以重载为类的友元函数。

1. 运算符重载为非静态成员函数

运算符重载为非静态成员函数的原型：

```
<返回值类型> operator<运算符>(<形式参数列表>);
```

其中，<返回值类型>可以是任何有效类型，由运算符的运算结果确定。operator 是运算符重载的关键词，C++中任何运算符重载函数名都必须使用此关键词。<运算符>表示要重载的运算符。<形式参数列表>包含了运算符所需的操作数，参数个数与运算符的操作数个数相关。每个非静态成员函数都有一个隐含的自引用参数 this 指针，对于单目运算符重载函数，不需要显式声明操作数，形式参数列表为空，所需的形参将由自引用参数提供。对于双目运算符重载函数，只需要显式声明右操作数，左操作数由自引用参数提供。在使用运算符做运算时，运算符的左操作数必须为类对象，否则会引起编译错误。因为使用运算符时仍然是函数调用，成员函数只有绑定到具体对象才能运行。例如，假设定义了一个复数类(CComplex)对象的实部加上一个实数的加法运算符重载函数，则有：

```
CComplex  a(1,1);              //定义复数类对象
double x=2.0;
cout<< a+x<<endl;              //对:实际转换为a.operator+(x)
cout<<x+a<<endl;               //错:变量x没有成员函数，x.operator+(a)无法匹配
```

2. 运算符重载为友元函数

运算符重载为友元函数的原型：

```
friend <返回值类型> operator<运算符>(<形式参数列表>);
```

其中，关键词 friend 用于声明运算符重载函数与类之间的友元关系，friend 关键词只能在函数声明位置出现，不在函数定义位置出现。<返回值类型>、<运算符>和<形式参数列表>的含义与重载为非静态成员函数相同。因为友元函数不是类的成员函数，没有隐含的自引用参数 this 指针，所以运算符重载为友元函数时，形式参数都必须显式声明。对于一元运算符重载函数，需要显式声明操作数，形式参数个数为 1 个。对于二元运算符重载函数，需要显式声明左操作数和右操作数，形参个数为 2 个。即运算符重载为友元函数时，函数参数个数与运算符操作数个数相同。

运算符重载为友元函数时，由于其参数个数与运算符操作数个数相同，所以直观上更容易理解运算符重载的原理，而且友元函数对类特有的访问权限保证了重载函数对数据操作的方便性。

运算符重载也可以重载为普通函数，其函数原型与重载为友元函数的原型相同，只是在声明函数没有表明重载函数与类关系的 friend 关键词。由于普通函数对类的访问权限限制了重载函数对类对象数据成员的操作，所以重载为普通函数对类对象的操作都必须通过成员函数完成，所以一般不将运算符重载为普通函数形式。

4.3 运算符重载规则

运算符重载可以使程序更加简洁，使表达更加直观，增加了程序的可读性，但运算符重载必须遵循以下规则。

1. 不能创建新的操作符

C++中运算符重载只能对 C++已经定义的运算符集合里的运算符进行重新定义，不允许创建新的运算符，如**不是 C++运算符，不允许将**重载为一个新的运算符以实现某种运算。

2. 运算符重载的限制

C++中的运算符集合包含有很多运算符，并不是所有的运算符都允许重载。C++中允许重载的运算符如下：

+	–	*	/	%	^	&
|	~	!	=	<	>	+=
–=	*=	/=	%=	^=	&=	|=
<<	>>	>>=	<<=	==	!=	<=
>=	&&	||	++	——	->*	,
->	[]	()	new	new[]	delete	delete[]

下面的运算符不允许重载：

::—域作用运算符

.—成员运算符

.*—指针成员的成员运算符

这 3 个运算符的第 2 个参数都要求是名称，而不是值，而且提供的是引用成员的最基本语义。允许对它们重载将导致难以琢磨的问题。唯一的三元运算符——条件运算符也不允许对其重载，此外 sizeof 和 typeof 也不允许重载。

1. 不能改变运算符的优先级和结合性

C++的运算符都具有特定的优先级和结合性，重载运算符后，运算符的优先级和结合性不能改变。如不可能通过运算符重载使得加法运算符优先级高于乘法运算符。

2. 不能改变操作数个数

运算符重载不能改变运算符的操作数个数，一元运算符重载后仍然是一元运算符，二元运算符重载后依然是二元运算符。运算符重载不改变运算符的功能，只是增加了运算符操作的对象类型以及针对该类型实现运算符功能的流程。

3. 不要改变运算符的含义

运算符重载就是对运算符赋予了操作自定义类对象的能力，但这种能力的赋予不应该改变运算符操作的本质，否则就会产生莫名其妙的结果。例如：如果将复数类的乘法重载为加法操作，加法操作重载为减法操作，那么在实际使用时会导致使用者完全无法理解的结果。例如：

```
CComplex  a, b;
a = a*b;           //实际上做 a+b
b = a+b;           //实际上做 a-b
```

对于任何没有仔细阅读过运算符重载函数原码的用户来说，永远也无法理解这样的运算符为什么会得到如此错误的结果。

4. 某些运算符重载的特殊要求

C++的部分运算符由于其操作本质只允许被重载为非静态成员函数，这类运算符为：

[]：下标运算符

=：赋值运算符

()：调用运算符

–>：指向运算符

因为以上 4 个运算符的操作必须通过类对象调用才能运行，所以这 4 个运算符只能重载为类的非静态成员函数。

有 2 个运算符只能重载为类的友元函数，分别是：

>>：输入运算符

<<：输出运算符

因为输入输出运算符只能通过输入流对象和输出流对象才能调用运行，不满足成员函数必须通过类对象调用才能运行的要求，所以只能重载为友元函数。

4.4 单目运算符重载

单目运算符是那些只有一个操作数的运算符，C++中比较常见的单目运算符有自增自减运算符。本节将以实例形式详细介绍单目运算符中自增自减运算符的重载方法。为了便于理解，本节以比较简单而又常用的复数类来介绍运算符重载的实现方法。

自增运算符有前自增和后自增两种，重载增量运算操作符时，虽然运算符(函数名)相同，但功能不同，应该有前自增和后自增的差别。这种差别应该体现在正确匹配函数的参数和反映操作本质的返回值两方面。

前自增操作的结果应该与对象一致，而且前自增操作的结果可以作为左值，操作可以连贯。后自增操作的结果是自增操作之前的对象，是一个临时对象，当表达式计算完成后，该对象就消失了，所以后自增返回值与后自增之后的对象是不一致的。例如：

```
int ia=1,ib=1,ic=1,id=1;
++(++ia);    //ia=3
(++ib)++;    //ib=3
(ic++)++;    //ic=2,(ic++)的结果是临时变量，其自增1后就被抛弃
++(id++);    //id=2,原因与ic相似
```

从上面代码可以看出，前自增运算符的返回值与对象是同一个值，使得对象可以连续地进行自增操作，所以前自增运算符重载函数的参数和返回值都必须是对象的引用。前自增运算符重载函数形式如下：

```
<类型&> operator++(<类型&>);    //函数原型：重载为友元函数
++ca;                            //等价于 ca.operator++(ca)
++(++ca);    //等价于 ca.operator++(ca)，返回对象 ca，再次 ca.operator++(ca)
```

后自增运算符重载函数需要改变原来的对象，函数参数必须是对象的引用，但函数返回一个临时对象，所以返回值不能是对象的引用，只能是对象值。另外，因为前自增运算符重载函数和后自增运算符重载函数的函数名相同，参数都是对象的引用，使得在函数调用时无法正确区分两个函数。所以，为了区别两种运算符重载函数，C++要求后自增运算符重载函数必须增加一个 int 型参数，但该参数在函数中不起任何作用，只用于区分前自增和后自增运算符重载函数。后自增运算符重载形式如下所示：

```
<类型> operator++(<类型&>, int);    //重载为友元函数
ca++;                                //等价于 ca.operator++(ca,1);
```

下面以一个复数类为例演示自增运算符重载函数的实现原理与使用。

```
/***************************************************************
    File name     : f0402.h
    Description   : 复数类头文件
***************************************************************/
#ifndef _COMPLEX_H
#define _COMPLEX_H
/*class define ----------------------------------------------*/
class CComplex
{
private:
    double m_im, m_re;
public:
    CComplex(double = 0, double = 0);        //构造函数
    void set_re(double);                     //设置实部的函数
    void set_im(double);                     //设置虚部的函数
```

```cpp
    void display()const;                              //常成员函数
    friend CComplex& operator++(CComplex&);     //前自增运算符重载函数
    friend CComplex operator++(CComplex&, int); //后自增运算符重载函数
};
#endif
//=============================================================
/****************************************************************
    File name     :  f0402.cpp
    Description   :  复数类源文件
****************************************************************/
#include<iostream>
#include"f0402.h"
usingnamespace std;
/*member function ------------------------------------------------*/
CComplex::CComplex(double re, double im)
{
    m_re = re;
    m_im = im;
}
inline void CComplex::set_re(double re)
{
    m_re = re;
}
inline void CComplex::set_im(double im)
{
    m_im = im;
}
void CComplex::display()const
{
    cout << m_re <<"+"<< m_im <<"i";
}
/*friend function------------------------------------------------*/
Ccomplex & operator++(CComplex& cobj)
{
    ++cobj.m_re;
    return cobj;
}
CComplex operator++(CComplex& cobj, int)
{
    CComplex temp(cobj);        //先保存原来对象到temp中
    ++cobj.m_re;
    return temp;
}
//=============================================================
/****************************************************************
    File name     :  f0402_driver.cpp
    Description   :  复数类测试程序
****************************************************************/
```

```cpp
#include<iostream>
#include"f0402.h"
usingnamespace std;
int main()
{
    CComplex complex1(1, 2), complex2;
    cout <<"complex1: ";
    complex1.display();             //显示complex1
    cout <<"\n 前自增 complex1: ";
    complex2 = ++complex1;
    complex1.display();
    cout <<"\n complex2: ";
    complex2.display();
    cout <<"\n 后自增 complex1:";
    complex2 = complex1++;
    complex1.display();
    cout <<"\n complex2: ";
    complex2.display();
    cout <<"\n";
    return 0;
}
//=============================================================
```

程序的运行结果为：

```
complex1: 1+2i
前自增 complex1: 2+2i
complex2: 2+2i
后自增 complex1: 3+2i
complex2: 2+2i;
```

注意，在 f0402.h 头文件开始和尾部的宏定义：

```
#ifndef _COMPLEX_H
#define _COMPLEX_H
    //……
#endif
```

该宏定义称为头文件卫士，主要是为了防止在多文件程序中，避免头文件重复包含引起的错误。当编译器第一次包含 f0402.h 头文件时，程序中没有定义宏_COMPLEX_H，则定义宏_COMPLEX_H，并包含由此宏定义包围的所有语句。如果有文件再次包含 f0402.h，因为程序中已经定义了宏_COMPLEX_H，就不会再次包含头文件中的所有语句。

上面复数类中的自增运算符重载为复数实部加 1，并将自增运算符重载为友元函数的形式。自增运算符也可以重载为成员函数。重载为成员函数时，函数参数是由 this 指针隐性声明，不需要显式声明形参。具体代码如下所示：

```cpp
class CComplex
{
private:
    double m_im, m_re;
```

```
public:
    //..........
    CComplex& operator++();        //前自增运算符重载函数
    CComplex operator++(int);      //后自增运算符重载函数
};
```

成员函数实现如下：

```
CComplex& CComplex::operator++()
{
    ++(this->m_re);
    return *this;
}
CComplex CComplex::operator++(int)
{
    CComplex temp(*this);          //先保存原来对象到 temp 中
    ++(this->m_re);
    return *this;
}
```

重载为成员函数时，其参数由 this 指针隐性声明，所以函数不需要参数。

自减运算符的重载方式与自增运算符相似，只是重载函数实现减 1 运算，读者可自行实现。

4.5 双目运算符重载

数学运算符、关系运算符、逻辑运算符、下标运算符等都是双目运算符，重载这些运算符可以使得程序在运算等方面更加直观。大部分双目运算符都可以重载为友元函数和成员函数，只有赋值运算符、下标运算符必须重载为成员函数。下面选择下标运算符、赋值运算符和加法运算符来讲述双目运算符重载原理以及实现方式，其他双目运算符重载原理与此类似。

4.5.1 下标运算符重载

下标运算符必须重载为非静态成员函数。下标运算符的功能是提取类对象的某个成员数据，提取的数据既可以作为右值使用，也可以作为左值使用，当作为左值使用表示该数据可以被修改。例如，数组的下标运算符：

```
int  i_arry[100], ia;
i_arry[0] = 10;                //下标运算符重载函数返回结果作为左值
ia = i_arry[0];                //下标运算符重载函数返回结果作为右值
```

如果下标运算符重载函数作为左值，则函数的返回值就必须是某个可以修改的对象。一般来说，函数的返回值不可以作为左值使用，因为它是一个值。除非一种例外的情况，就是函数返回的是某个对象的地址或引用。如果返回某个对象的地址，则作为左值使用时，需要取值运算才能表示该对象。例如：

```
int  i_arry[100];
*i_arry[0] = 10;               //当下标运算符返回对象地址时的情况，显然不符合常规
```

因为下标运算符重载函数返回指针，所以必须进行取值运算才能获取其指向的对象。这种使用显然不合理，因为数组元素表示一个个整型变量，在整型变量上进行取值运算符显然

不合理。因此，**下标运算符重载函数只能返回对象的引用**。下标运算符重载函数的原型：

```
<类型&> operator[](int);
```

其中，<类型&>表示下标运算符所属类对象的引用，表示返回某个对象的引用。int 表示下标的值。

下面以一个字符串类来演示下标运算符重载函数的使用。

```cpp
/******************************************************************
    File name    : f0403.h
    Description  : 字符串类头文件
******************************************************************/
#ifndef _MYSTRING_H
#define _MYSTRING_H
/*class define------------------------------------------------*/
class CMyString
{
private:
    unsigned int m_size;            //字符串的最大长度-1
    char *mp_data;                  //指向存放数据的内存空间
public:
    CMyString(unsignedint = 0);     //构造函数
    CMyString(char* pstr);          //构造函数
    ~CMyString();                   //析构函数
    char & operator[](int);         //下标运算符重载函数
    int lenght()const;              //返回当前数组中元素个数
    int capacity()const;            //返回数组的最大元素个数
    void display()const;            //显示字符串函数
};
#endif
//================================================================
/******************************************************************
    File name    : f0403.cpp
    Description  : 字符串类源文件
******************************************************************/
#include<iostream>
#include"f0402.h"
usingnamespace std;
/*member function define--------------------------------------*/
CMyString::CMyString(unsigned int size)
{
    m_size = size;
    if (size > 0)
    {
        mp_data = new char[size];
    }
    else
    {
        mp_data = 0;                //将指针赋值为空
    }
```

```cpp
}
CMyString::CMyString(char* pstr)
{
    m_size = strlen(pstr) + 20;         //为字符串类对象预留20个字符空间
    mp_data = new char[m_size];         //申请内存空间
    strcpy(mp_data, pstr);              //复制字符串内容到mp_data
}
CMyString::~CMyString()
{
    if (mp_data != 0)
        delete[] mp_data;
}
inline int CMyString::lenght()const
{
    return strlen(mp_data);
}
inline int CMyString::capacity()const
{
    return m_size;
}
char& CMyString::operator[](int index)
{
    return mp_data[index];
}
void CMyString::display()const
{
    cout << mp_data << endl;
}
//================================================================
/******************************************************************
    File name    :   f0403_driver.cpp
    Description  :   字符串类测试程序
******************************************************************/
#include<iostream>
#include"f0403.h"
usingnamespace std;
int main()
{
    CMyString str("ABCDEFG");                   //定义一个字符串对象

    cout <<"Orignal string:"<< endl;
    str.display();
    str[1] = '5';                               //下标运算符作为左值
    cout <<"Modified string:"<< endl;
    str.display();
    cout <<"No.1 char: "<< str[1] << endl;      //下标运算符作为右值
    return 0;
}
//================================================================
```

程序的输出结果为：

```
Orignal string:
ABCDEFG
Modified string:
A5CDEFG
5
```

4.5.2 赋值运算符重载

如果设计类时没有自定义赋值运算符重载函数，则系统会为类提供一个默认的赋值运算符重载函数。如果自定义了赋值运算符重载函数，则系统不再提供默认赋值运算符重载函数。系统提供的默认赋值运算符重载函数只能实现类数据成员的按位赋值，是一种浅复制形式的赋值运算操作。例如：

```
CComplex complex1(1,2), complex2;
complex2 = complex1;
```

前面程序 f0402 定义的 complex 类没有定义赋值运算符重载函数，则使用系统提供的默认赋值运算符重载函数完成赋值运算操作。其赋值过程如图 4-1 所示。

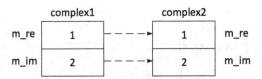

图 4-1 浅复制赋值运算符的赋值过程

利用 f0403 中 CMyString 写出下列语句：

```
CMyString str1("ABCDE"),str2(10);
str2 = str1;
```

同样，前面定义的 CMyString 类没有定义赋值运算符重载函数，则上面的赋值语句使用系统提供的默认赋值运算符重载函数完成赋值操作。其过程如图 4-2 所示。

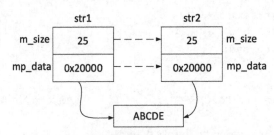

图 4-2 浅复制赋值运算符的赋值结果

从上面的赋值过程可以看到，字符串对象 str1 和 str2 的数据指针指向了同一个字符串，使得两个对象共享了相同的数据。如果程序中某个对象消失了，则导致另外一个对象的数据指针指向无效的内存，从而产生无法预料的结果或者异常。例如：

```
CMyString *pstr1=new CMyString("ABCDE"), str2(10);
str2 = *pstr1;          //调用系统提供的赋值运算符重载函数
delete pstr1;           //释放 pstr1 所指向的对象。
```

上面语句中当释放了 pstr1 所指对象后，其数据成员 mp_data 所指的数据也同时被释放了，这使得 str2 的指针成员 mp_data 指向一个无效的内存空间，当再使用该内存时就会产生异常或错误。

仔细对比 CComplex 类和 CMyString 类的差异，可以发现 CComplex 类的对象实体和对象本体是一致的，而 CMyString 类有指针成员，使得其对象实体和对象本体不一致。

特别提醒：如果类的对象本体和对象实体不一致，则必须为类提供自定义赋值运算符重载函数。

一般来说，当类有指针成员时，必须为类自定义赋值运算符重载函数，以实现对象的深复制。

下面为 CMyString 类添加赋值运算符重载函数，类的定义修改为如下形式，增加了赋值运算符重载函数。

```cpp
/******************************************************************
    File name    : f0404.h
    Description  : 字符串类头文件
******************************************************************/
#ifndef _MYSTRING_H
#define _MYSTRING_H
/*class define--------------------------------------------------*/
class CMyString
{
private:
    unsignedint m_size;            //字符串的最大长度-1
    char *mp_data;                 //指向存放数据的内存空间
public:
    CMyString(unsignedint = 0);    //构造函数
    CMyString(char* pstr);         //构造函数
    ~CMyString();                  //析构函数
    int size() const;              //返回当前数组中元素个数
    int lenght()const;             //返回当前数组中元素个数
    int capacity()const;           //返回数组的最大元素个数
    void display()const;           //显示字符串函数
    char& operator[](int);         //下标运算符重载函数
    CMyString& operator=(const CMyString&);    //赋值运算符重载
};
#endif
//================================================================
```

CMyString 类赋值运算符重载函数实现如下：

```cpp
/*member function----------------------------------------------*/
CMyString& CMyString::operator=(const CMyString& obj)   //赋值运算符重载
{
    if (this == &obj)                 //判断两个对象是否相同
    {
        return *this;
    }
```

```
    m_size = obj.m_size;
    if (mp_data != 0)                      //如果mp_data不为0，则释放该内存
    {
        delete[]mp_data;
    }
    mp_data = new char[obj.m_size];        //根据对象大小重新申请内存
    strcpy(mp_data, obj.mp_data);          //复制数据到新申请的内存空间
    return *this;
}
```

在赋值运算符重载函数中，第3~6行判断是否是对象自己给自己赋值，如果时，则直接返回对象本身；如果不是，才进行真正的赋值操作。在赋值过程中，要注意释放对象原来占有的内存空间，否则会引起内存泄漏问题。

4.5.3 加法运算符重载

加法运算符和其他数学运算符是计算机程序中最常用的数学运算符。加法运算法即可以重载为友元函数，又可重载为成员函数。下面以复数类为例演示加法运算符重载的实现方式。

复数类CComplex的定义修改如下：

```
/*****************************************************************
    File name   : f0405.h
    Description : 复数类头文件
*****************************************************************/
#ifndef _COMPLEX_H
#define _COMPLEX_H
/*class define ------------------------------------------------*/
class CComplex
{
private:
    double m_im, m_re;
public:
    CComplex(double = 0, double = 0);      //构造函数
    void set_re(double);                   //设置实部的函数
    void set_im(double);                   //设置虚部的函数
    void display()const;                   //常成员函数
    friend CComplex&operator++(CComplex&);              //前自增运算符重载函数
    friend CComplex operator++(CComplex&, int);         //后自增运算符重载函数
    friend CComplex operator+(const CComplex&, const CComplex&); //加法运算符
};
#endif
//==============================================================
```

上面的复数类CComplex定义中，将加法运算符重载为友元函数，加法运算符重载函数的定义如下：

```
/*friend function----------------------------------------------*/
CComplex operator+(const CComplex& c1, const CComplex& c2)
{
    CComplex temp(c1);       //用c1复制构造临时对象temp
```

```
            temp.m_re += c2.m_re;
            temp.m_im += c2.m_im;
            return temp;
        }
```

上面两个复数相加的加法运算符重载函数定义中,首先用参数中的复数对象构造一个临时对象,然后再将临时对象与另外一个参数进行加法运算,返回临时对象。重载函数以类对象的常引用为函数参数,提高了函数参数传递的效率,同时通过 const 限定来保护实参值不被重载函数修改。加法运算符的使用方法如下:

```
    CComplex  complex1(1,2), complex2(3,4), complex3;
    complex3 = complex1 + complex2;    //两个复数对象相加
```

上面第 2 行语句执行之后,**complex3** 为 4+6i。

加法运算符也可以重载为成员函数,此时只需显式声明一个形参,另一个形参由 this 指针隐性声明。函数原型为:

```
    CComplex operator+(const CComplex&);
```

函数定义如下:

```
    CComplex CComplex::operator+(const CComplex& c)
    {
        this->m_re += c.m_re;
        this->m_im += c.m_im;
        return *this;
    }
```

4.6 流插入运算符和流提取运算符的重载

因为流插入运算符和流提取运算符的第一个操作数必须是流对象,所以这两个运算符只能重载为友元函数。

流提取运算符重载函数的结构如下:

```
    istream& operator>>(istream& in, classtype& obj)
    {
        //输入操作
        return in;
    }
```

流插入运算符重载函数的结构如下:

```
    ostream& operator<<(ostream& out, const classtype& obj)
    {
        //输出操作
        return out;
    }
```

流提取运算符重载函数的第 1 个参数为输入流对象,它既可以是文件输入流,又可以是标准输入设备流。第 2 参数表示保存提取数据的对象,必须是引用作为参数。流插入运算符重载函数的第 1 个参数为输出流对象,既可以是文件输出流,也可以是标准输出设备流。第 2

个参数表示要输出的对象，可以是对象的引用，也可以是对象的常引用，但建议使用对象的常引用作为函数参数。因为引用参数只需传递一个指针值，而传值形式需要传递整个对象，所以引用作为参数可以提高函数参数的传递效率。常引用可以保护引用参数不被重载函数修改。

下面仍以复数类为例来演示流插入运算符和流提取运算符重载函数的实现方式和使用方法。

复数类 CComplex 的完整定义如下：

```
/******************************************************************
    File name    : f0406.h
    Description  : 复数类头文件
******************************************************************/
#ifndef _COMPLEX_H
#define _COMPLEX_H
/*class define ---------------------------------------------------*/
class CComplex
{
private:
    double m_im, m_re;
public:
    CComplex(double = 0, double = 0);        //构造函数
    void set_re(double);                     //设置实部的函数
    void set_im(double);                     //设置虚部的函数
    void display()const;                     //常成员函数
    friend CComplex&operator++(CComplex&);   //前自增运算符重载函数
    friend CComplex operator++(CComplex&, int);  //后自增运算符重载函数
    friend CComplex operator+(const CComplex&, const CComplex&);
                                             //加法运算符
    friend istream& operator>>(istream&, CComplex&);   //流提取运算符
    friend ostream& operator<<(ostream&, CComplex&);   //流插入运算符
};
#endif
//================================================================
```

完整的复数类 CComplex 成员函数实现如下：

```
/******************************************************************
    File name    : f0406.cpp
    Description  : 复数类源文件
******************************************************************/
#include<iostream>
#include"f0406.h"
using namespace std;
/*member function ------------------------------------------------*/
CComplex::CComplex(double re, double im)
{
    m_re = re;
    m_im = im;
}
inline void CComplex::set_re(double re)
{
```

```cpp
        m_re = re;
    }
    inline void CComplex::set_im(double im)
    {
        m_im = im;
    }
    void CComplex::display()const
    {
        cout << m_re <<"+"<< m_im <<"i";
    }
    /*friend function-------------------------------------------------------*/
    CComplex& operator++(CComplex& cobj)
    {
        ++cobj.m_re;
        return cobj;
    }
    CComplex operator++(CComplex& cobj, int)
    {
        CComplex temp(cobj);         //先保存原来对象到 temp 中
        ++cobj.m_re;
        return temp;
    }
    CComplex operator+(const CComplex& c1, const CComplex& c2)
    {
        CComplex temp(c1);           //用 c1 复制构造临时对象 temp
        temp.m_re += c2.m_re;
        temp.m_im += c2.m_im;
        return temp;
    }
    istream& operator>>(istream& in, CComplex& obj)
    {
        in >> obj.m_re >> obj.m_im;
        return in;
    }
    ostream& operator<<(ostream& out, CComplex& obj)
    {
        out << obj.m_re <<"+"<< obj.m_im <<"i";
        return out;
    }
    //============================================================
```

完整的复数类测试程序如下：

```cpp
    /*****************************************************************
        File name   : f0406_driver.cpp
        Description : 复数类测试文件
    *****************************************************************/
    #include<iostream>
    #include"f0406.h"
```

```cpp
using namespace std;
int main()
{
    CComplex complex1, complex2, complex3;
    cout <<"请输入一个复数: "<< endl;
    cin >> complex1;
    cout <<"complex1: "<< complex1 << endl;        //显示 complex1
    cout <<"前自增 complex1:";
    complex2 = ++complex1;
    cout << complex1 << endl;
    cout <<"前自增 complex1 的结果:";
    cout << complex2 << endl;
    cout <<"后自增 complex1:";
    complex2 = complex1++;
    cout << complex1 << endl;
    cout <<"后自增 complex1 的结果: ";
    cout << complex2 << endl;
    complex3 = complex1 + complex2;
    cout <<"complex1 + complex2 : "<< complex3 << endl;
    complex3 = complex1 + 2.1;
    cout <<"complex1+2.1: "<< complex3 << endl;
    complex3 = 2.5 + complex1;
    cout <<"2.5+complex1: "<< complex3 << endl;
    return 0;
}
//===============================================================
```

程序的输出结果为:

```
请输入一个复数:
1 2
complex1: 1+2i
前自增 complex1: 2+2i
前自增 complex1 的结果: 2+2i
后自增 complex1:3+2i
后自增 complex1 的结果: 2+2i
complex1 + complex2 : 5+4i
complex1+2.1: 5.1+2i
2.5+complex1: 5.5+2i
```

4.7 不同类型数据间的转换

类是用户自定义的类型，类之间、类和基本类型之间都可以像系统预定义的基本类型之间一样进行类型转换，实现这种转换要使用类型转换构造函数和类类型转换运算符重载函数。

4.7.1 基本数据类型到类类型的转换

对于 4.5 节中带有加法运算符重载函数的复数类 CComplex，假设有如下语句:

```
    CComplex c1(1,3), c2;
    c2 = c1+4;
    c2 = 4+c1;
```

语句第 2 和第 3 行都能正常运行,并且能得到正确结果。观察文件 f0405.h 中 CComplex 类的定义中,可以看到 CComplex 类定义的加法运算符重载函数的参数是两个 CComplex 类对象,而上面语句第 2 和第 3 行的加法表达式中只有一个操作数是 CComplex 对象,这说明系统对程序做了一种类型转换,也就是将整型数 4 转换成了一个临时的复数对象。那么,系统又是通过什么途径完成这个转换的呢?

C++中,只要定义的类中存在带有一个参数的构造函数,则允许通过该构造函数将参数类型对象转换为一个类对象。上列中的 CComplex 类有一个带两个默认参数值的构造函数,语句第 2 和 3 行中表达式为了匹配加法运算符重载函数,系统将 4 通过带默认参数值的构造函数转换成了临时对象 4+0i,从而使得复数对象能够与 4 相加。这种将其他类型数据转换成类对象的构造函数称为转型构造函数。

类型转换构造函数的推导必须遵守以下规则:
(1) 类型转换构造函数必须是只有一个参数的构造函数。
(2) 类型转换只推导一次,如果遇到二义性,则放弃推导。例如:

```
/****************************************************************
    File name   :  f0407.cpp
    Description :  基本数据类型到类类型的转换
****************************************************************/
#include<iostream>
usingnamespace std;
/*class define------------------------*/
class CMoney
{
private:
    int m_yuan, m_jiao, m_fen;
public:
    CMoney(double);                    //double 类型转换为 CMoney 类型函数
    CMoney(int, int, int);
    void display()const;
};
/*member function--------------------*/
CMoney::CMoney(int y, int j, int f)
{
    m_yuan = y;
    m_jiao = j;
    m_fen = f;
}
CMoney::CMoney(double x)
{
    m_yuan = (int)x;
    m_jiao = (int)(x * 10) - m_yuan * 10;
    m_fen = (int)(x * 100) - m_yuan * 100 - m_jiao * 10;
}
```

```cpp
void CMoney::display()const
{
    cout << m_yuan <<"元"<< m_jiao <<"角"<< m_fen <<"分";
}

void print(CMoney& money)
{
    money.display();
    cout << endl;
}
void main()
{
    print(2.13);
}
//========================================
```

上面程序中，第 38 行 print 函数无法将 double 数据 2.13 转换成 CMoney 类对象，需要做显式类型转换才能正确匹配，应该修改为以下形式。

```cpp
print(CMoney(2.13));
```

4.7.2 类类型到基本类型的转换

类型转换构造函数实现了基本类型到类类型之间的转换，但有时也需要从类类型到基本类型之间的转换。C++中引入了一种特殊的成员函数称为类型转换函数，这种类类型转换函数实际上是类类型转换运算符重载函数。

类类型转换函数是专门用来将类类型转换为基本数据类型，只能被重载为成员函数。类类型转换函数格式：

```
operator <返回类型名>()
{
    //……
    return  <基本类型值>
}
```

例如：

```cpp
/****************************************************************
    File name   :  f0408.cpp
    Description :  类类型到其他类型的转换
****************************************************************/
#include<iostream>
usingnamespace std;
class CMoney
{
private:
    int m_yuan, m_jiao, m_fen;
public:
    CMoney(double);              //double 类型转换为 CMoney 类型函数
    CMoney(int, int, int);
    operator double();           //CMoney 类型转换为 double 类型函数
```

```
    void display()const;
};
/*member function--------------------*/
CMoney::CMoney(int y, int j, int f)
{
    m_yuan = y;
    m_jiao = j;
    m_fen = f;
}
CMoney::CMoney(double x)
{
    m_yuan = (int)x;
    m_jiao = (int)(x * 10) - m_yuan * 10;
    m_fen = (int)(x * 100) - m_yuan * 100 - m_jiao * 10;
}
void CMoney::display()const
{
    cout << m_yuan <<"元"<< m_jiao <<"角"<< m_fen <<"分";
}
CMoney::operator double()
{
    double value = m_yuan + m_jiao*0.1 + m_fen * 0.01;
    return value;
}
/*测试程序----------------------------*/
int main()
{
    CMoney money(1, 2, 3);
    cout << money << endl;
    return 0;
}
//========================================
```

程序的输出结果为：

```
1.23
```

有了类类型转换函数之后，就可以将 CMoney 对象转换成 double 数据。

类型转换运算符重载函数的特点：

(1) 类类型转换运算符重载函数没有返回值类型，<返回类型名>就代表返回值类型，但函数体中必须有 return 语句来返回相应的值。

(2) 类类型转换运算符重载函数没有参数，但在调用过程中必须有一个对象实参。

当类的一个对象为表达式中的一个操作数时，而表达式要求该操作数应该具有<返回值类型名>要求的类型时，系统将自动调用类型转换函数，将类对象转换为相应类型。

4.8 综合应用实例

日期和时间在任何计算机系统中都是非常重要的一个数据，做任何事情都应该记录该事情发生的日期和时间。比如：从数据库里读取一条数据、修改数据库的一条数据、从图书馆借一本书等。下面将通过定义一个日期类 CDate 和一个时间类 CTime,以及一个表示日期和时间的类 CDateTime 来演示运算符重载功能。

CDate 类提供以下功能:
(1)带默认参数值的构造函数,用于定义一个默认的日期对象,值为 2000-01-01。
(2)带一个字符串 string 对象的构造函数,通过字符串构造一个日期对象。
(3)带一个 C 字符串的构造函数,用于通过 C 字符串构造一个日期对象。
(4)设置日期函数。
(5)加法运算符重载函数,实现日期加多少天的功能,需要考虑月和年的进位。
(6)减法运算符重载函数,用于计算两个日期相差多少天。
(7)前自增/后自增运算符重载函数,实现日期自动加一天的功能,需要考虑月和年的进位。
(8)流提取运算重载函数。
(9)流插入运算符重载函数。
(10)类型转换运算符重载函数,将日期类对象转换成 string 字符串。

CTime 类提供如下功能:
(1)默认参数值的构造函数,用于定义一个默认时间对象,值为 0,0,0。
(2)带一个字符串 string 对象的构造函数,通过字符串构造一个时间对象。
(3)带一个 C 字符串的构造函数,用于通过 C 字符串构造一个时间对象。
(4)设置时间函数。
(5)加法运算符重载函数,实现时间增加多少秒,需要考虑分和时的进位。
(6)减法运算符重载函数,用于计算两个时间相差多少小时多少分多少秒。
(7)前自增/后自增运算符重载函数,实现时间自动增加一秒的功能,需要考虑时分的进位。
(8)流提取运算符重载函数。
(9)流插入运算符重载函数。
(10)类型转换运算符重载函数,将时间对象转换成 string 字符串。

CDateTime 类提供如下功能:
(1)默认构造函数,用于定义一个默认的日期时间对象,值为 2000-01-01;00:00:00。
(2)带两个字符串 string 对象的构造函数,通过字符串构造一个日期时间对象。
(3)带两个 C 字符串的构造函数,通过 C 字符串构造一个日期时间对象。
(4)设置日期函数。
(5)设置时间函数。
(6)流提取运算符重载函数。
(7)流插入运算符重载函数。

程序代码如下:

```
/****************************************************************
    File name    :  datetime.h
    Description  :  日期时间类头文件
****************************************************************/
#ifndef _DATE_TIME_H
#define _DATE_TIME_H

#include <iostream>
```

```cpp
#include <string>

/*class CDate define ---------------------------------------------*/
class CDate
{
private:
    int m_year, m_month, m_day;              //年月日
public:
    CDate(int = 2000, int = 1, int = 1);     //构造函数
    CDate(const std::string&);               //构造函数
    CDate(const char*);                      //构造函数
    void set_date(int, int, int);
    int  IsLeap()const;                      //常成员函数
    /*friend function --------------------------------------*/
    friend CDate operator+(CDate, int);      //加法
    friend int operator-(CDate, CDate);
    friend CDate& operator++(CDate&);        //前自增
    friend CDate operator++(CDate&, int);    //后自增
    friend std::istream& operator>>(std::istream&, CDate&); //流提取运算符
    friend std::ostream& operator<<(std::ostream&, const CDate&);
                                             //流插入运算符
};
/*class CTime define ---------------------------------------------*/
class CTime
{
private:
    int m_hour, m_minute, m_second;
public:
    CTime(int = 0, int = 0, int = 0);        //构造函数
    CTime(const std::string&);               //构造函数
    CTime(const char*);
    void set_time(int, int, int);

    /*friend function --------------------------------------*/
    friend CTime operator+(CTime, int);      //加法
    friend int operator-(CTime, CTime);      //减法
    friend CTime& operator++(CTime&);        //前自增
    friend CTime operator++(CTime&, int);    //后自增
    friend std::istream& operator>>(std::istream&, CTime&);
                                             //流提取运算符
    friend std::ostream& operator<<(std::ostream&, const CTime&);
                                             //流插入运算符
};
/*class CDateTime define -----------------------------------------*/
class CDateTime
{
private:
    CDate m_date;
```

```cpp
        CTime m_time;
    public:
        CDateTime(int = 2000, int = 01, int = 01, int = 0, int = 0, int = 0);
        CDateTime(const CDate&, const CTime&);
        void set_date(int, int, int );         //设置日期函数
        void set_time(int, int, int );         //设置时间函数
        void add_date(int);                    //增加天数
        void add_time(int);                    //增加秒数

        /*friend function ----------------------------------------*/
        friend std::istream& operator>>(std::istream&, CDateTime&);
                                               //流提取运算符
        friend std::ostream& operator<<(std::ostream&, const CDateTime&);
                                               //流插入运算符
};
#endif
//======================================================================

/************************************************************************
    File name     : datetime.cpp
    Description   : 日期时间类源文件
************************************************************************/
#include <iostream>
#include <string>
#include "datetime.h"
using namespace std;

/*globle const define----------------------------------------*/
const int month_days[] = {0, 31, 28, 31, 30, 31, 30, 31, 31, 30, 31, 30, 31};
/*member function of class CDate ----------------------------*/
CDate::CDate(int year, int month, int day)
{
    m_year = year;
    m_month = month;
    m_day = day;
}
CDate::CDate(const string& str)
{
    int index = 0;

    index = str.find("-", 0);
    string str_temp = str.substr(0, index);
    m_year = atoi(str_temp.c_str());
    int index2 = str.find("-", index + 1);
    str_temp = str.substr(index + 1, index2 - index - 1);
    m_month = atoi(str_temp.c_str());
    str_temp = str.substr(index2 + 1, str.length() - index2);
    m_day = atoi(str_temp.c_str());
```

```cpp
}
CDate::CDate(const char* str)
{
    string str_bak = str;
    int index = 0;

    index = str_bak.find("-", 0);
    string str_temp = str_bak.substr(0, index);
    m_year = atoi(str_temp.c_str());
    int index2 = str_bak.find("-", index + 1);
    str_temp = str_bak.substr(index + 1, index2 - index - 1);
    m_month = atoi(str_temp.c_str());
    str_temp = str_bak.substr(index2 + 1, str_bak.length() - index2);
    m_day = atoi(str_temp.c_str());

}
void CDate::set_date(int year, int month, int day)
{
    m_year = year;
    m_month = month;
    m_day = day;
}
int CDate::IsLeap()const
{
    if ((m_year % 400 == 0) || ((m_year % 4 == 0) && (m_year % 100 != 0)))
        return 1;
    else
        return 0;
}
/*friend function ----------------------------------*/
/*此函数为全局内联函数，用于判断某年是否为闰年 ---------*/
inline bool IsLeap(int year)
{
    return ((year % 400 == 0) || ((year % 4 == 0) && (year % 100 != 0)));
}

CDate operator+(CDate date, int num)
{
    date.m_day += num;
    /*处理日的进位--------*/
    if ((2 == date.m_month) && (date.IsLeap()))   //处理2月问题
    {
        if (date.m_day > month_days[date.m_month] + 1)
        {
            date.m_day -= month_days[date.m_month] + 1;
            ++date.m_month;
        }
```

```cpp
        }
        else
        {
            if (date.m_day > month_days[date.m_month])
            {
                date.m_day -= month_days[date.m_month];
                ++date.m_month;
            }
        }
        /*处理月的进位---------*/
        if (date.m_month > 12)
        {
            date.m_month -= 12;
            ++date.m_year;
        }
        return date;
}
int operator-(CDate date1, CDate date2)
{
        int sum_days = 0, y_days = 0, m_days = 0;
        int flag = 0; //0--date1>=date2; 1--date1<date2
        CDate temp;

        /*先排序日期，保证date1>=date2 -------------*/
        if ((date1.m_year < date2.m_year)
            || ((date1.m_year == date2.m_year) && (date1.m_month < date2.m_month))
            || ((date1.m_year == date2.m_year) && (date1.m_month == date2.m_month)
                && (date1.m_day < date2.m_day)))
        {
            temp = date1;
            date1 = date2;
            date2 = temp;
            flag = 1;
        }

        int i = date2.m_year + 1;
        //计算年间差的天数
        while (i < date1.m_year)
        {
            y_days += IsLeap(i) ? 366 : 365;
            ++i;
        }
        //计算date1 和 date2 本身的天数
        int value1 = 0;
        for (i = 0; i < date1.m_month; ++i)
        {
            value1 += month_days[i];
        }
```

```cpp
        value1 += date1.m_day;
        if (date1.IsLeap())
        {
            value1 += date1.m_month>2 ? 1 : 0;
        }
        int value2 = 0;
        for (i = 0; i < date2.m_month; ++i)
        {
            value2 += month_days[i];
        }
        value2 += date2.m_day;
        if (date2.IsLeap())
        {
            value1 += date2.m_month>2 ? 1 : 0;
        }
        if (date1.m_year == date2.m_year)
        {
            m_days = value1 - value2;
        }
        else
        {
            m_days = (date2.IsLeap()) ? (366 - value2 + value1) : (365 - value2 + value1);
        }

        return flag == 0 ? (y_days + m_days) : -(y_days + m_days);
    }

    CDate& operator++(CDate& date)                    //前自增
    {
        date = date + 1;
        return date;
    }
    CDate operator++(CDate& date, int)                //后自增
    {
        CDate temp(date);
        date = date + 1;
        return temp;
    }
    std::istream& operator>>(std::istream& in, CDate& date)    //流提取运算符
    {
        in >> date.m_year >> date.m_month >> date.m_day;
        return in;
    }

    std::ostream& operator<<(std::ostream& out, const CDate& date)
                                                            //流插入运算符
    {
        out << date.m_year << "-" << date.m_month << "-" << date.m_day;
```

```cpp
        return out;
}

/*member function of class CTime --------------------------------*/
CTime::CTime(int hour, int minute, int second)         //构造函数
{
    m_hour = hour;
    m_minute = minute;
    m_second = second;
}

CTime::CTime(const string& str)                        //构造函数
{
    int start_index;
    string str_temp;

    start_index = str.find(":", 0);
    str_temp = str.substr(0, start_index);
    m_hour = atoi(str_temp.c_str());
    int index = str.find(":", start_index + 1);
    str_temp = str.substr(start_index + 1, index - start_index);
    m_minute = atoi(str_temp.c_str());
    str_temp = str.substr(index + 1, str.length() - index);
    m_second = atoi(str_temp.c_str());
}
CTime::CTime(const char* pstr)
{
    string str;
    str = pstr;

    int start_index;
    string str_temp;

    start_index = str.find(":", 0);
    str_temp = str.substr(0, start_index);
    m_hour = atoi(str_temp.c_str());
    int index = str.find(":", start_index + 1);
    str_temp = str.substr(start_index + 1, index - start_index);
    m_minute = atoi(str_temp.c_str());
    str_temp = str.substr(index + 1, str.length() - index);
    m_second = atoi(str_temp.c_str());
}
void CTime::set_time(int hour, int minute, int second)
{
    m_hour = hour;
    m_minute = minute;
    m_second = second;
}
```

```cpp
/*friend function ----------------------------------------*/
CTime operator+(CTime time, int num)            //加法
{
    time.m_second += num;
    if (time.m_second >= 60)                    //秒进位处理
    {
        time.m_second -= 60;
        ++time.m_minute;
        if (time.m_minute >= 60)                //分钟进位处理
        {
            time.m_minute -= 60;
            ++time.m_hour;
            time.m_hour %= 24;                  //小时进位处理
        }
    }
    return time;
}
int operator-(CTime time1, CTime time2)         //减法
{
    int sum_second, hour_second, minute_second;
    int flag = 0; //0--time1>=time2;1--time1<time2
    CTime temp;

    //先保证 tiem1>time2
    if ((time1.m_hour < time2.m_hour)
        || ((time1.m_hour == time2.m_hour)&&(time1.m_minute<time2.m_minute))
        || ((time1.m_hour == time2.m_hour) && (time1.m_minute==time2.m_minute)
            && (time1.m_second<time2.m_second)))
    {
        temp = time1;
        time1 = time2;
        time2 = temp;
        flag = 1;
    }
    //计算小时的秒数
    hour_second = (time1.m_hour - time2.m_hour - 1) * 3600;
    minute_second = (60 - time2.m_minute) * 60 + time1.m_minute * 60;
    sum_second = hour_second + minute_second - time2.m_second + time1.m_second;
    return flag==0?sum_second:-sum_second;
}
CTime& operator++(CTime& time)                  //前自增
{
    time = time + 1;
    return time;
}
CTime operator++(CTime& time, int)              //后自增
{
```

```cpp
        CTime temp(time);
        time = time + 1;
        return temp;
    }
    std::istream& operator>>(std::istream& in, CTime& time)     //流提取运算符
    {
        in >> time.m_hour >> time.m_minute >> time.m_second;
        return in;
    }
    std::ostream& operator<<(std::ostream& out, const CTime& time)//流插入运算符
    {
        out << time.m_hour << ":" << time.m_minute << ":" << time.m_second;
        return out;
    }
    /*member function of class CDateTime ----------------------------*/
    CDateTime::CDateTime(int year, int month, int day, int hour, int minute, int second)
            :m_date(year, month, day), m_time(hour, minute, second)
    {
    }
    CDateTime::CDateTime(const CDate& date, const CTime& time) : m_date(date), m_time(time)
    {
    }
    void CDateTime::set_date(int year, int month, int day)   //设置日期函数
    {
        m_date.set_date(year, month, day);
    }
    void CDateTime::set_time(int hour, int minute, int second)    //设置时间函数
    {
        m_time.set_time(hour, minute, second);
    }
    void CDateTime::add_date(int days)                  //增加天数
    {
        m_date = m_date + days;
    }
    void CDateTime::add_time(int seconds)               //增加秒数
    {
        m_time = m_time + seconds;
    }

    /*friend function --------------------------------------*/
    std::istream& operator>>(std::istream& in, CDateTime& date_time)
                                                        //流提取运算符
    {
        in >> date_time.m_date >> date_time.m_time;
        return in;
    }
    std::ostream& operator<<(std::ostream& out, const CDateTime& date_time)
                                                        //流插入运算符
```

```cpp
{
    out << date_time.m_date << ":" << date_time.m_time;
    return out;
}
//================================================================

/******************************************************************
    File name   : datetime-driver.h
    Description : 日期时间类测试程序文件
******************************************************************/
#include <iostream>
#include "datetime.h"
using namespace std;

int main()
{
    CDate date1, date2("2015-08-26");
    cout << "date1: " << date1 << endl;
    cout << "date2: " << date2 << endl;
    CDate date3 = date1 + 52;
    cout << "date3: " << date3 << endl;
    int num = date1- date3;
    cout << "num: " << num << endl;
    cout << "date1++: " << date1++ << endl;
    cout << "date1: " << date1 << endl;
    cout << "++date1: " << ++date1 << endl;
    cout << "date1: " << date1 << endl;

    cout << "==============================\n";

    CTime time1, time2("9:23:34");
    cout << "time1: " << time1 << endl;
    cout << "time2: " << time2 << endl;
    CTime time3 = time1 + 82;
    cout << "time3: " << time3 << endl;
    num = time2 - time1;
    cout << "num: " << num << endl;
    cout << "time1++: " << time1++ << endl;
    cout << "time1: " << time1 << endl;
    cout << "++time1: " << ++time1 << endl;
    cout << "time1: " << time1 << endl;

    cout << "==============================\n";

    CDateTime datetime1, datetime2(2015, 8, 27, 10, 45, 35);
    cout << "datetime1: " << datetime1 << endl;
    cout << "datetime2: " << datetime2 << endl;
    datetime1.add_date(3);
```

```
        cout << "datetime1: " << datetime1 << endl;
        datetime1.add_time(100);
        cout << "datetime1: " << datetime1 << endl;
        datetime1.set_date(2010, 8, 27);
        cout << "datetime1: " << datetime1 << endl;
        datetime1.set_time(10, 48, 21);
        cout << "datetime1: " << datetime1 << endl;

        return 0;
    }
```

程序的输出结果为：

```
    date1: 2000-1-1
    date2: 2015-8-26
    date3: 2000-2-22
    num: -52
    date1++: 2000-1-1
    date1: 2000-1-2
    ++date1: 2000-1-3
    date1: 2000-1-3
    ==============================
    time1: 0:0:0
    time2: 9:23:34
    time3: 0:1:22
    num: 33814
    time1++: 0:0:0
    time1: 0:0:1
    ++time1: 0:0:2
    time1: 0:0:2
    ==============================
    datetime1: 2000-1-1:0:0:0
    datetime2: 2015-8-27:10:45:35
    datetime1: 2000-1-4:0:0:0
    datetime1: 2000-1-4:0:1:40
    datetime1: 2010-8-27:0:1:40
    datetime1: 2010-8-27:10:48:21
```

练 习

1. 定义一个二维平面的点类CPoint，包含两个坐标属性，请完成如下功能：
(1)定义一个带默认参数值的构造函数，参数默认值都为0。
(2)定义一个前自增运算符重载函数，要求实现CPoint对象的两个坐标分别加1。
(3)定义一个后自增运算符重载函数，要求实现CPoint对象的两个坐标分别加1。
(4)定义一个加法运算符重载函数，要求实现CPoint对象的两个坐标分别加上一个实数。
(5)定义一个流提取运算符重载函数。
(6)定义一个流插入运算符重载函数。

(7)编程 CPoint 类的测试程序。

2. 设计一个书本类 CBook，该类包含书名、作者名、出版社名、价格、出版日期、书本数量等属性，其中书本数量表示某本书名的图书数量。请完成如下功能：

(1)定义一个日期类，并使用日期类对象作为 CBook 类的成员。

(2)定义一个带默认参数值的构造函数，其中参数默认值：书名为 BookName；作者名为 AuthorName；出版社名为 PressName；价格为 0.0；出版日期为 2015-1-1。

(3)定义一个流提取运算符重载函数。

(4)定义一个流插入运算符重载函数。

(5)定义修改书本数量的成员函数，注意该成员函数的特性，建议将书本数量属性设置为静态数据成员，该函数定义为静态成员函数。

(6)定义一个加法运算符重载函数，实现书本价格加上一个实数的功能。

(7)请自行增加其他成员函数

(8)编写 CBook 类的测试程序。

3. 在上题的基础上定义一个图书管理类 CBookMng，该类管理的图书数量没有限制，按要求完成如下功能：

(1)自行设计 CBookMng 类的数据成员。

(2)定义 CBookMng 类的构造函数，使得在定义 CBookMng 类对象时，可以给定初始值，也可以不给初始化。

(3)定义一个复制构造函数。

(4)定义一个析构函数。

(5)定义一个加法函数，实现两个 CBookMng 类对象的合并运算。

(6)定义一个流插入运算符重载函数，用于将 CBookMng 类对象管理的所有图书信息输出到屏幕上，要求每本图书输出一行。

(7)定义一个下标运算符重载函数，用于检索下标指定的图书信息，要求考虑下标越界异常。

(8)自行确定增加成员函数。

(9)编写 CBookMng 类的测试程序。

4. 向量是一种很重要的数据类型，表达了一个相同类型数据的有序集合。很多物理量如力、速度、位移以及电场强度、磁感应强度等都是向量。试着定义一个向量类型，使得向量类型具有如下性质和功能：

(1)向量元素个数没有限制，请思考如何实现。

(2)向量对象可分为行向量和列向量，提供向量的转置操作，行向量转置变为列向量，列向量转置变为行向量

(3)相同维数的向量加法、减法，向量的加法和减法仍然是一个向量，结果向量的每个元素为两个向量对应元素的和或差。

(4)相同维数的向量内积，行向量乘以列向量结果为一个标量(单个值)。

(5)获取向量的维数。

(6)获取向量的最大容量。

(7)获取向量的元素个数。

(8)向量的下标操作获取向量的一个元素。

(9)计算向量的模。

第 5 章 继承与派生

通过前面章节的学习，可以方便地创建简单对象。但面向对象程序设计不仅仅如此，还进一步地扩展面向对象技术，可以提供类型和子类型的关系。这是通过一种被称为继承的机制获得的。类不仅重新实现共享的特征，也继承了其父类的数据成员和成员函数。C++通过一种被称为类派生的机制来支持继承，被继承的类为基类，而新的类称为派生类。把基类和派生类实例的集合称为类继承层次结构。

- **知识要点**

  理解面向对象技术的可重用性；

  掌握面向对象程序设计的重要特性之一——继承性，及其使用方法。

- **探索思考**

  在 C 语言中，软件的重用方法有哪些形式？

  有了继承关系后，为什么还要进一步讨论面向对象的另一大特性——多态性？

- **预备知识**

  复习类类型的设计，特别是类对数据成员和成员函数的封装和访问控制。

5.1 继承与派生的概念

1. 继承与类

继承(Inheritance)是指一个事物可以继承其父辈全部或部分特性，同时本身还可以有自己的特性。C++将继承的概念应用于类的机制中，但类的继承和自然界的继承概念有些不一样。自然界的继承遵循的是遗传规律，而 C++类的继承使用的是**概括**(Generalization)的方法。

可以这么理解，通过继承，C++能够定义这样的类类型，它们对类型之间的关系建模，共享公共的东西，仅仅是特化本质上不同的东西。子类能够继承父类的成员，子类只需要重新定义那些与子类类型相关的成员函数。因此，子类除了从父类继承的成员之外，还可以有子类自己的成员。

2. 类继承相关概念

子类继承父类，也可以称为父类派生了子类。从已定义的类派生了的新类，不仅继承了原有类的属性和方法，并且还可以拥有自己新的属性和方法，这就是类的**继承**和**派生**。被继承的类也称为**基类**(Base class)、**超类**(Super class)或者**父类**，在基类或父类基础上建立的新类称为**派生类**(Derived class)或**子类**(Subclass)。

3. 类层次关系的描述方法

上述父类和子类的关系称为**类层次**，或称为**继承关系**。在类设计时，常常将这些关系用**树**来描述。例如，用树来描述学校人员的各类层次关系。

CStaff 类（职员类）是在 CPerson 类（人员类）的基础上建立的，因此 CPerson 类称为 CStaff 类的基类，CStaff 类称为 CPerson 类的派生类。若基类 CPerson 类描述的属性有姓名、出生年月、性别、电话、身份证号码等，那么派生类 CStaff 类就可以继承 CPerson 类的这些属性，并且还可以有自己的属性，如入校工作时间、调入方式、工作经历、学历、学位等。除了 CStaff 类是 CPerson 类的派生类外，CStudent 类（学生类）也是 CPerson 类的派生类。可见，**一个基类可以有多个派生类**。并且，CAdministrator 类（管理人员）和 CTeacher 类（教师类）又是 CPerson 类的派生类 CStaff 的派生类，同样，CGraduateStudent 类（研究生）是 CPerson 类的派生类 CStudent 的派生类。因此，一个派生类还可以作为基类，继续派生出新的类来，这样的派生方式称为**多层派生**，从继承的角度来看称为**多层继承**，如图 5-1 所示。

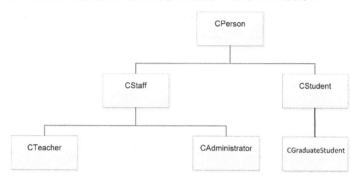

图 5-1 继承关系

5.2 定义基类和派生类

基类和派生类的定义在许多方面像我们已经见过的其他类定义一样。但是，在继承层次中定义类还需要另外一些特性，本节和后续的章节将逐步介绍这些特性以及如何使用基类和派生类来编写程序。

5.2.1 定义基类

基类定义基本的**数据成员**和**成员函数**。在简化的学生应用程序的例子中，CStudent 类定义了 name（学生姓名）、semesterHours（总修学时）和 average（平均分）数据成员，以及 addCourse（添加选课）、display（显示信息）、getHours（获取总修学时）和 getAverage（获取平均分）函数：

```
/*************************************************************
    File name    : f0501.cpp
    Description  : 定义学生基类
*************************************************************/
#include <iostream>
#include <string>
using namespace std;

class CStudent
{
protected:
```

```cpp
        string name;
    private:
        int semesterHours;
        double average;
    public:
        CStudent(string pName="noName")
            :name(pName),average(0),semesterHours(0)          //初始化参数列表
        {
        }
        void addCourse(int hours, double grade)
        {
            double totalGrade = (semesterHours * average + grade*hours);   //总分
            semesterHours+=hours;                                           //总修学时
            average=semesterHours ?totalGrade/semesterHours:0;  //平均分
        }
        void display()
        {
            cout<<"name=\""<<name<<"\""<<", hours="<<semesterHours<<",
                < average="<<average<<"\n";
        }
        int getHours()
        {
            return semesterHours;
        }
        double getAverage()
        {
            return average;
        }
        virtual ~CStudent()
        {
        }
};
int main()
{
    CStudent s;
    s.display();
    return 0;
}

//======================================
```

程序的输出结果为:

```
name="noName", hours=0, average=0
```

这个类定义了一个**构造函数**以及已描述过的函数，该构造函数使用**默认实参**，允许用 0 个或 1 个实参进行调用，通过定义的**初始化参数列表**，使用这些实参初始化数据成员。初始化的次序就是定义成员的次序。第 1 个成员首先被初始化，然后是第 2 个，依次类推。与**初始化参数列表**的列表顺序无关。

通过初始化列表来初始化成员，这是一个好的习惯，如果将上面例子中的构造函数改写成这样：

```cpp
/******************************************************************
    File name   : f0502.cpp
    Description : 定义学生基类
******************************************************************/
#include <iostream>
#include <string>
using namespace std;
class CStudent
{
protected:
    string name;
private:
    int semesterHours;
    double average;
public:
    CStudent(string pName="noName")
    {
        name=pName;
        average=0;
        semesterHours=0;
    }
    void addCourse(int hours, double grade)
    {
        double totalGrade = (semesterHours * average + grade * hours);  //总分
        semesterHours+=hours;                                            //总修学时
        average=semesterHours ?totalGrade/semesterHours:0; //平均分
    }
    void display()
    {
        cout<<"name=\""<<name<<"\""<<", hours="<<semesterHours<<",
                average="<<average<<"\n";
    }
    int getHours()
    {
        return semesterHours;
    }
    double getAverage()
    {
        return average;
    }
    virtual ~CStudent()
    {
    }
};
int main()
{
    CStudent s;
    s.display();
    return 0;
```

```
            }

            //======================================
```

程序的输出结果为：

```
            name="noName", hours=0, average=0
```

程序的运行结果与例 f0501.cpp 的运行结果是一样的，但实际上，这种初始化的方法在初始化 name、average 和 semesterHours 的时机落后于初始化参数列表，这是由 C++的规则决定的。

特别提醒：初始化的次序常常无关紧要。然而，如果一个成员初始化受到其他成员初始化的影响，则成员初始化的次序至关重要。因此，按照与成员声明一致的次序编写构造函数初始化列表是个好习惯。此外，通过初始化参数列表指定初始化对象成员是对象初始化的好办法。

1. 基类成员函数

CStudent 类定义了两种函数，其中一种在其返回类型前面带有保留字 virtual，另一种没有。有保留字 virtual 的为虚函数，不带保留字 virtual 的则不是虚函数。除了构造函数之外，任意非 static 成员函数都可以是虚函数。保留字只在类内部的成员函数声明中出现，不能在类定义体外部出现的函数定义中。关于虚函数，将在后面章节进一步讨论。

2. 访问控制和继承

在基类中，public 和 private 关键字具有普通含义：用户代码可以访问类的 public 成员而不能访问 private 成员，private 成员只能由基类的成员和友元访问。派生类对基类的 public 和 private 成员的访问权限与程序中任意其他部分一样：它可以访问 public 成员而不能访问 private 成员。

有时作为基类的一些成员，它希望允许派生类访问，同时禁止其他用户访问。对于这样的成员应使用受保护的访问关键字 protected。protected 成员（受保护的成员）可以被派生类对象访问但不能被该类型的普通用户访问。

我们的 CStudent 类希望它的派生类重定义 display 函数，为了重定义 display 函数，这些类将需要访问 semesterHours 和 average 成员。希望派生类用户与普通用户一样通过 getHours 和 getAverage 成员函数访问 private 的 semesterHours 和 average 数据成员，它们不能被 CStudent 的继承类所访问；而 protected 的 name 可以被 CStudent 的继承类所访问。

5.2.2 定义派生类

为了定义派生类，使用类派生列表指定基类。类派生列表指定了一个或多个基类，声明派生类的一般形式为：

```
            class 派生类名：[继承方式]基类名列表
            {
                派生类新增加的成员
            };
```

继承方式包括：public（公用的）、private（私有的）和 protected（受保护的），此项是可选的，如果不选此项，则默认为 private（私有的），基类名列表是已定义的基类名的列表，列表项以逗号分割。也就是说派生类可以指定多个基类。继承单个基类最为常见，称为单继承，否则称为多重继承，下面分别讨论这两种继承方式。

继承方式决定了对继承成员的访问权限。如果想要继承基类的接口，则应该选择 public 派生方式。

派生类继承基类的成员并且可以定义自己的附加成员。每个派生类对象包含两个部分：从基类继承的成员和自己定义的成员。一般而言，派生类只重新定义那些与基类不同的，或者扩展基类属性和方法。

1. 派生类构成

在学生应用程序中，将从 CStudent 类(学生类)派生 CGraduateStudent 类(研究生类)，因此 CGraduateStudent 类将继承 name、semesterHours 和 average 数据成员和 addCourse、display、getHours 和 getAverage 函数成员。派生类 CGraduateStudent 重新定义了 display 成员函数，在派生类中的 display 成员函数覆盖了基类 display 成员函数的操作方法。同时，定义了自己的 getQualifier 成员函数。

```cpp
class CGraduateStudent: public CStudent
{
    CAdvisor advisor;
    int qualifierGrade;
public:
    int getQualifier()
    {
        return qualifierGrade;
    }
    void display()
    {
        cout<<"name=\""<<name<<"\""<<", hours="<<getHours()<<",
         average="<<getAverage()<<", qualifierGrade ="<< qualifierGrade <<"\n";
    }
};
```

2. 派生类对象包含基类对象作为子对象

派生类对象由多个部分组成：派生类本身定义的(非静态)成员加上由基类(非静态)成员组成的子对象。CStudent 对象就是由两个部分组成的。

派生类对象的定义方法：

```cpp
CGraduateStudent gs;
```

gs 为 CStudent 的派生类 CStudent 的对象，其组成部分为图 5-2 所示。

| 基类CStudent对象部分 |
| 派生类CGraduateStudent对象部分 |

图 5-2 派生类对象的组成

派生对象虽然是由基类部分和派生类部分组成，但只表示这两个部分逻辑上是连续排列的，但物理存储角度上不一定是连续排列的。

3. 派生类中的函数可以使用基类的成员

派生类的成员函数可以在类的内部或外部定义,如这里的 getQualifier 和 display 函数一样：

```cpp
class CGraduateStudent : public CStudent
{
    CAdvisor advisor;
```

```cpp
        int qualifierGrade;
    public:
        int getQualifier()
        {
            return qualifierGrade;
        }
        void display()
        {
            cout<<"name=\""<<name<<"\""<<", hours="<<getHours()<<",
            average="<<getAverage()<<", qualifierGrade ="<< qualifierGrade <<"\n";
        }
    };
```

或者在类中声明，类外定义

```cpp
    class CGraduateStudent: public CStudent
    {
        CAdvisor advisor;
        int qualifierGrade;
    public:
        int getQualifier();
        void display();
    };

    int CGraduateStudent ::getQualifier()
    {
        return qualifierGrade;
    }
    void CGraduateStudent ::display()
    {
        cout<<"name=\""<<name<<"\""<<", hours="<<getHours()
        <<", average="<<getAverage()<<", qualifierGrade ="<< qualifierGrade <<"\n";
    }
```

因为每个派生类对象都有基类部分，类中可以访问该基类对象的 **public** 和 **protected** 成员，就好像那些成员是派生类对象自己的成员一样。

4. 用作基类的类必须是已定义的

已定义的类才可以用作基类。如果已经声明了 CStudent 类，但没有定义它，则不能用作基类：

```cpp
    class CStudent;                     //没有定义基类
    class CGraduateStudent : public CStudent
    {
        ...
    };
```

这一限制的原因在于：每个派生类包含并且可以访问其基类的成员，为了使用这些成员，派生类必须知道它们是什么。

在定义派生类时，一般还应当自己定义派生类的构造函数和析构函数。

C++中的构造函数和析构函数是不能从基类继承的。需要在派生类中定义自己的构造函数和析构函数。如果没有定义，编译器将会为派生类提供的默认构造函数和析构函数。

5. 继承和派生实例分析

实际上定义基类和派生类时，可以先定义一个基类，在此基类中只提供某些最基本的功能，然后在声明派生类时加入某些具体的功能，形成适用于某一特定应用的派生类。通过对基类定义的延续，将一个基类转化成具体的派生类。因此，派生类是基类的具体化版本。

用继承关系来描述动物世界的特征和关系：

(1) 抽象出项目问题中的对象：动物、老鼠、熊猫。

(2) 抽象出每种对象所具有的性质：名字和食物。

(3) 抽象出每种对象所具有的行为，动物具有吃的行为和睡觉的行为，老鼠除具有动物行为外还具有打洞行为。

(4) 抽象出继承关系，老鼠类和熊猫类作为子类继承父类动物类，子类继承父类的属性和方法。

描述动物世界的特征和关系：

(1) 动物世界的名字和食物是共有的属性，吃和睡觉是共有的行为。定义 Animal 类，在该类中定义成员变量 name 和 food，成员方法 eat()和 sleep()。根据封装性的要求将 name 和 food 属性定义为私有的，保证该属性只能被在本类内部访问，其他类或派生类要访问该属性可以通过 public 公用的方法 setName()、getName()、setFood()、getFood()进行访问，其中 setXXX()方法用于给属性赋值，getXXX()方法用于返回属性值。

(2) 定义老鼠类 Mouse 和熊猫类 Panda，这两类均继承自动物类 Animal，继承动物类 Animal 的 name 和 food 属性以及 eat()和 sleep()方法。Mouse 类和 Panda 类在构造方法中调用 setName()和 setFood()方法对名称和食物属性初始化。

(3) 老鼠类定义打洞方法 dig()输出信息。

(4) 定义测试类 AnimalTest，编写程序入口 main()方法，在该方法中创建老鼠和熊猫对象，调用相应方法输出信息。

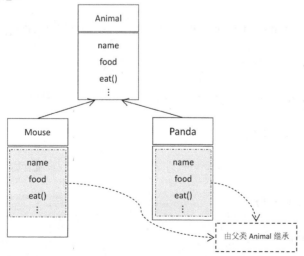

图 5-3 用继承关系来描述动物世界的特征和关系(一)

图 5-3 表明了 Mouse 和 Panda 与 Animal 之间的继承和派生关系。父类 Animal 中的成员可以派生给子类 Mouse 和子类 Panda，因此子类中就不再需要额外定义它们。图 5-4 就是这种简化的表达。

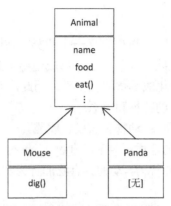

图 5-4 用继承关系来描述动物世界的特征和关系(二)

```
/*****************************************************************
    File name   : f0503.h
    Description : 定义动物 animal 基类、Mouse 派生类和 Panda 派生类
*****************************************************************/
#include <iostream>
using namespace std;
class Animal                              //动物类
{
private:
    string name;                          //名字
    string food;                          //食物
public:
    string getName()                      //返回名字
    {
        return name;
    }
    void setName(string nam)              //设置名字
    {
        name = nam;
    }
    string getFood()                      //返回食物
    {
        return food;
    }
    void setFood(string f)                //设置食物
    {
        food = f;
    }
    void eat()                            //吃
    {
        cout<<"我在吃饭\t";
```

```cpp
    }
    void sleeping()                                //睡觉
    {
        cout<<"我在睡觉\t";
    }
    Animal(string nam,string f)                    //构造方法
    {
        name = nam;
        food = f;
    }
};
class Mouse :public Animal                         //老鼠类
{
public:
    Mouse(string nam,string f):Animal(nam,f)       //构造方法
    {
    }
    void dig()                                     //打洞
    {
        cout<<"我在打洞\t";
    }
};
class Panda :public Animal                         //熊猫类
{
public:
    Panda(string nam,string f) :Animal(nam,f)      //构造方法
    {
    }
};

/******************************************************************
    File name   :  f0503.cpp
    Description :  动物类测试主程序
******************************************************************/
#include <iostream>
#include <string>
#include "f0503.h"
using namespace std;
void main()
{
    Panda panda_1 ("盼盼","竹叶");                  //实例化熊猫
    cout<<"名字："<<panda_1.getName()<<"\t";
    cout<<"食物："<<panda_1.getFood()<<"\t";
    panda_1.eat();                                  //吃
    panda_1.sleeping();                             //睡觉
    cout<<endl;
    Mouse mouse_1("米奇","大米");                   //实例化老鼠
    cout<<"名字："<<mouse_1.getName()<<"\t";
```

```
            cout<<"食物："<<mouse_1.getFood()<<"\t";
            mouse_1.eat();                              //吃
            mouse_1.sleeping();                         //睡觉
            mouse_1.dig();                              //打洞
        }

        //======================================
```

程序的输出结果为：

名字：盼盼食物：竹叶我在吃饭我在睡觉
名字：米奇食物：大米我在吃饭我在睡觉我在打洞

5.3 基类成员在派生类中的可访问性

派生类继承了基类的成员，但在派生类中对基类成员的访问由两个因素决定：成员在基类中的访问属性和继承时使用何种继承方式。

一个类的 public 成员对任何类和函数都是可见的，而 private 成员只对本类或本类的友元可见，对其他类和函数都是不可见的，对派生类亦是如此。一个类的 protected 成员的访问属性介于 public 和 private 之间：对它的派生类来说，基类的 protected 成员是可见的，但对其他类而言，protected 成员就如同 private 成员一样不可访问。既可实现某些成员的隐蔽，又可方便地继承。

继承方式也有三种访问方式：公用继承、私有继承和受保护继承方式。

根据成员在基类中的访问权限和继承时使用的何种继承方式，将基类不同访问属性的成员在三种不同继承方式下派生类的访问属性如表 5-1 所示。

表 5-1 基类成员在派生类中的访问属性

基类中的成员	在公用派生类中的访问属性	在私有派生类中的访问属性	在保护派生类中的访问属性
私有成员	不可访问	不可访问	不可访问
公用成员	公用	私有	保护
保护成员	保护	私有	保护

5.3.1 公用继承

在定义一个派生类时将基类的继承方式指定为 public 的，称为**公用继承**，用公用继承方式建立的派生类称为**公用派生类**(public derived class)，其基类称为**公用基类**(public base class)。

采用公用继承方式时，基类的公用成员和保护成员在派生类中仍然保持其公用成员和保护成员的属性，而基类的私有成员在派生类中并没有成为派生类的私有成员，它仍然是基类的私有成员，只有基类的成员函数可以引用它，而不能被派生类的成员函数引用，因此在派生类中是不可访问的。

```
        /************************************************************
            File name    : f0504.cpp
            Description  : CStudent 公用派生 CGraduateStudent
        ************************************************************/
        #include <iostream>
```

```cpp
#include <string>
using namespace std;
class CStudent                                    //声明基类
{
public :                                          //基类公用成员
    CStudent()
    {
        num=0;
        name="noName";
        sex='f';
    }
    void get_value()
    {
        cin>>num>>name>>sex;
    }
    void display()
    {
        cout<<" num: "<<num<<endl;
        cout<<" name: "<<name<<endl;
        cout<<" sex: "<<sex<<endl;
    }

private :                                         //基类私有成员
    int num;
    string name;
    char sex;
};

class CGraduateStudent: public Cstudent //以public方式声明派生类CGraduateStudent
{
public:
    CGraduateStudent()
    {
        age=0;
        addr="empty";
    }
    void display_GraduateStudent ()
    {
        cout<<" num: "<<num<<endl;           //企图引用基类的私有成员，错误
        cout<<" name: "<<name<<endl;         //企图引用基类的私有成员，错误
        cout<<" sex: "<<sex<<endl;           //企图引用基类的私有成员，错误
        cout<<" age: "<<age<<endl;           //引用派生类的私有成员，正确
        cout<<" address: "<<addr<<endl;      //引用派生类的私有成员，正确
    }
private:
    int age;
    string addr;
};
//=====================================
```

程序由于在派生类中引用基类的私有成员,编译出现错误,不能运行。

由于基类的私有成员对派生类来说是不可访问的,因此在派生类中的 display_GraduateStudent 函数中直接引用基类的私有数据成员 num、name 和 sex 是不允许的。只能通过基类的公用成员函数来引用基类的私有数据成员。可以将派生类 CGraduateStudent 的声明改为:

```
class CGraduateStudent: public Cstudent
                                    //以public方式声明派生类CGraduateStudent
{
public:
    void display_GraduateStudent()
    {
        cout<<" age: "<<age<<endl;       //引用派生类的私有成员,正确
        cout<<" address: "<<addr<<endl;  //引用派生类的私有成员,正确
    }
private:
    int age;
    string addr;
};
```

然后在主函数中分别调用基类中的 display 函数和派生类中的 display_GraduateStudent 函数,输出 5 个数据。

可以这样写 main 函数(假设对象 stud 中已有数据):

```
int main()
{
    CGraduateStudent stud;              //定义派生类GraduateStudent的对象stud
    stud.display ();                    //调用基类公用成员函数,输出派生类中三个
                                        //数据成员的值
    stud.display_GraduateStudent ();    //调用派生类公用成员函数,输出派生类中
                                        //两个数据成员的值
    return 0;
}
```

修改 f0504.cpp 后的程序:

```
/*****************************************************************
File name   : f0505.cpp
Description : CStudent公用派生CGraduateStudent
*****************************************************************/
#include <iostream>
#include <string>
using namespace std;
class CStudent                          //声明基类
{
public :                                //基类公用成员
    CStudent()
    {
        num=0;
        name="noName";
```

```cpp
            sex='f';
        }
        void get_value()
        {
            cin>>num>>name>>sex;
        }
        void display()
        {
            cout<<" num: "<<num<<endl;
            cout<<" name: "<<name<<endl;
            cout<<" sex: "<<sex<<endl;
        }
    private:                                    //基类私有成员
        int num;
        string name;
        char sex;
};

class CGraduateStudent: public  Cstudent
                                    //以public方式声明派生类CGraduateStudent
{
public:
    CGraduateStudent()
    {
        age=0;
        addr="empty";
    }
    void display_GraduateStudent()
    {
        cout<<" age: "<<age<<endl;              //引用派生类的私有成员，正确
        cout<<" address: "<<addr<<endl;         //引用派生类的私有成员，正确
    }
private:
    int age;
    string addr;
};
int main()
{
    CGraduateStudent stud;          //定义派生类GraduateStudent的对象stud
    stud.display ();                //调用基类公用成员函数，输出派生类中三个数
                                    //据成员的值
    stud.display_GraduateStudent ();//调用派生类公用成员函数，输出派生类中两个
                                    //数据成员的值
    return 0;
}

//======================================
```

程序运行结果为：

```
num: 0
name: noName
sex: f
age: 0
address: empty
```

当然也可以这样去改写派生类中的 display_GraduateStudent 函数，

```cpp
class CGraduateStudent: public Cstudent
                                   //以 public 方式声明派生类 CGraduateStudent
{
public:
    void display_GraduateStudent()
    {
        display();                          //调用基类中的公用成员 display 函数，正确
        cout<<" age: "<<age<<endl;          //引用派生类的私有成员，正确
        cout<<" address: "<<addr<<endl;     //引用派生类的私有成员，正确
    }
private:
    int age;
    string addr;
};
```

那么，main 函数为：

```cpp
int main()
{
    CGraduateStudent stud;                  //定义派生类 GraduateStudent 的对象 stud
    stud.display_GraduateStudent();         //调用派生类公用成员函数
                                            //输出基类和派生类的五个数据成员的值
    return 0;
}
//===================================
```

运行结果与例 f0505.cpp 的运行结果相同。

5.3.2 私有继承

在声明一个派生类时将基类的继承方式指定为 private 的，称为**私有继承**，用私有继承方式建立的派生类称为**私有派生类**(private derived class)，其基类称为**私有基类**(private base class)。

私有基类的公用成员和保护成员在派生类中的访问属性相当于派生类中的私有成员，即派生类的成员函数能访问它们，而在派生类外不能访问它们。

私有基类的私有成员在派生类中仍然是不可访问的成员，只有基类的成员可以使用它们。

一个基类成员在基类中的访问属性和在派生类中的访问属性可能是不同的。

将 f0505.cpp 中的公用继承方式改为私有继承方式(基类 CStudent 不改变)。可以写出私有派生类如下：

```cpp
class CGraduateStudent: private Cstudent
                        //以private方式声明派生类CGraduateStudent
{
public:
    void display_GraduateStudent()
    {
        cout<<" age: "<<age<<endl;      //引用派生类的私有成员，正确
        cout<<" address: "<<addr<<endl; //引用派生类的私有成员，正确
    }
private:
    int age; string addr;
};
```

主函数这样改写：

```cpp
int main() {
    CGraduateStudent stud1;             //定义一个CGraduateStudent类的对象stud1
    stud1.display();                    //私有基类的公用成员函数在，错误
                                        //派生类中是私有函数，外界不能引用
    stud1.display_GraduateStudent ();   //Display_1 函数是Student1类的，正确
                                        //公用函数
    return 0;
}
```

完整程序为：

```cpp
/********************************************************************
    File name   : f0506.cpp
    Description : CStudent 私有派生 CGraduateStudent
********************************************************************/
#include <iostream>
#include <string>
using namespace std;
class CStudent                          //声明基类
{
public :                                //基类公用成员
    CStudent()
    {
        num=0;
        name="noName";
        sex='f';
    }
    void get_value()
    {
        cin>>num>>name>>sex;
    }
    void display()
    {
        cout<<" num: "<<num<<endl;
```

```cpp
            cout<<" name: "<<name<<endl;
            cout<<" sex: "<<sex<<endl;
        }
    private :    //基类私有成员
        int num;
        string name;
        char sex;
};

class CGraduateStudent: private Cstudent
                            //以private方式声明派生类CGraduateStudent
{
public:
    CGraduateStudent()
    {
        age=0;
        addr="empty";
    }
    void display_GraduateStudent()
    {
        cout<<" age: "<<age<<endl;          //引用派生类的私有成员，正确
        cout<<" address: "<<addr<<endl;     //引用派生类的私有成员，正确
    }
private:
    int age; string addr;
};

int main()
{
    CGraduateStudent stud1;     //定义一个CGraduateStudent类的对象stud1
    stud1.display();            //私有基类的公用成员函数在派生类中是私有函数，
                                //外界不能引用，错误
    stud1.display_GraduateStudent ();//Display_1函数是Student1类的公用函数,
                                //正确
    return 0;
}
//===================================
```

程序由于私有基类的公用成员函数在派生类中是私有函数，外界不能引用，因此编译出现错误，不能运行。

注意，在派生类外 stud1.display()虽然不能访问私有基类的公用成员函数，但派生类的成员函数可以访问私有基类的公用成员（如 stud1.display_GraduateStudent 函数可以调用基类的公用成员函数 display，但不能引用基类的私有成员 num 等），可以通过调用基类的公用成员函数进一步来访问私有基类的私有成员。

因此可以将程序 f0506.cpp 改造为：

```
/*****************************************************************
    File name : f0507.cpp
```

```
    Description :  CStudent 私有派生 CGraduateStudent
*********************************************************************/
#include <iostream>
#include <string>
using namespace std;
class CStudent                             //声明基类
{
public :                                   //基类公用成员
    CStudent()
    {
        num=0;
        name="noName";
        sex='f';
    }
    void get_value()
    {
        cin>>num>>name>>sex;
    }
    void display()
    {
        cout<<" num: "<<num<<endl;
        cout<<" name: "<<name<<endl;
        cout<<" sex: "<<sex<<endl;
    }
private :                                  //基类私有成员
    int num;
    string name;
    char sex;
};
class CGraduateStudent: private Cstudent
                         //以private方式声明派生类CGraduateStudent
{
public:
    CGraduateStudent()
    {
        age=0;
        addr="empty";
    }
    void display_GraduateStudent()
    {
        display();                        //调用基类中的公用成员display函数,正确
        cout<<" age: "<<age<<endl;        //引用派生类的私有成员,正确
        cout<<" address: "<<addr<<endl;   //引用派生类的私有成员,正确
    }
private:
    int age; string addr;
};
int main()
```

```
        {
            CGraduateStudent stud;              //定义派生类GraduateStudent的对象stud
            stud.display_GraduateStudent ();    //调用派生类公用成员函数
                                                //输出基类和派生类的五个数据成员的值
            return 0;
        }
    //======================================
```

程序运行结果为：

```
num: 0
name: noName
sex: f
age: 0
address: empty
```

这样就能正确地引用私有基类的私有成员。

由于私有继承对基类的行为和属性的继承限制很强，因此一般不使用该继承方式。

5.3.3 保护成员和保护继承

由 protected 声明的成员称为**受保护的成员**，或简称**保护成员**。从类的用户角度来看，保护成员等价于私有成员。但有一点与私有成员不同，保护成员可以被派生类的成员函数引用。

如果基类声明了私有成员，那么任何派生类都是不能访问它们的，若希望在派生类中能够访问，应当把它们声明为保护成员。如果在一个类中声明了保护成员，就意味着该类可能要用作基类，在它的派生类中会访问这些成员。

在定义一个派生类时将基类的继承方式指定为 protected 的，称为**保护继承**，用保护继承方式建立的派生类称为**保护派生类**(protected derived class)，其基类称为**保护基类**(protected base class)。

保护继承的特点：保护基类的公用成员和保护成员在派生类中都成了保护成员，其私有成员仍为基类私有，不能被派生类访问。也就是保护基类的所有成员在派生类中都被保护起来，类外不能访问，其公用成员和保护成员可以被其派生类的成员函数访问。

比较一下私有继承和保护继承，也就是比较在私有派生类中和在保护派生类中的访问属性，可以发现在直接派生类中，以上两种继承方式的作用实际上是相同的。在类外不能访问任何成员，而在派生类中可以通过成员函数访问基类中的公用成员和保护成员。但是如果继续派生，在新的派生类中，两种继承方式的作用就不同了。例如，如果以公用继承方式派生出一个新派生类，原来私有基类中的成员在新派生类中都成为不可访问的成员，无论在派生类内或外都不能访问，而原来保护基类中的公用成员和保护成员在新派生类中为保护成员，可以被新派生类的成员函数访问。

下面通过一个例子说明怎样访问保护成员。

```
    /*****************************************************************
        File name    : f0508.cpp
        Description  : 在派生类中引用保护成员
    *****************************************************************/
```

```cpp
#include <iostream>
#include <string>
using namespace std;
class CStudent                                      //声明基类
{
public :                                            //基类公用成员
    CStudent()
    {
        num=0;
        name="noName";
        sex='f';
    }
    void display();
protected:                                          //基类保护成员
    int num;
    string name;
    char sex;
};
void CStudent::display()                            //类外定义基类成员函数

{
    cout<<"num: "<<num<<endl;
    cout<<"name: "<<name<<endl;
    cout<<"sex: "<<sex<<endl;
}
class CGraduateStudent: protected  CStudent //用protected方式声明派生类
{
                                                    //CGraduateStudent
public :
    CGraduateStudent()
    {
        age=0;
        addr="empty";
    }
    void display_GraduateStudent ();                //派生类公用成员函数
private :
    int age;                                        //派生类私有数据成员
    string addr;                                    //派生类私有数据成员
};
void CGraduateStudent::display_GraduateStudent ()//定义派生类公用成员函数
{
    display();                                      //引用基类的公用成员函数,进一步
                                                    //访问基类的保护成员,合法
    cout<<"sex: "<<sex<<endl;                       //引用基类的保护成员,合法
    cout<<"age: "<<age<<endl;                       //引用派生类的私有成员,合法
    cout<<"address: "<<addr<<endl;                  //引用派生类的私有成员,合法
}
int main()
```

```
{
    CGraduateStudent stud1;                    //stud1 是派生类 CGraduateStudent
                                               //类的对象
    stud1.display_GraduateStudent ();          //display_GraduateStuden()是派生
                                               //类中的公用成员函数，合法
    stud1.num=10086;                           //外界不能访问保护成员错误
    return 0;
}
//======================================
```

程序由于外界不能访问保护成员，因此编译出现错误，不能运行。

由于在派生类的成员函数中引用基类的保护成员是合法的。为了修改出错程序，可以在派生类中增加一个成员函数 void CGraduateStudent::set_GraduateNum 函数：

```
class CGraduateStudent: protected  CStudent //用 protected 方式声明派生类
{
public :
    CGraduateStudent()
    {
        age=0;
        addr="empty";
    }
    void display_GraduateStudent ();                    //派生类公用成员函数
    void CGraduateStudent:: set_GraduateNum (int nNo); //添加派生类公用成员函数
private :
    int age;                                            //派生类私有数据成员
    string addr;                                        //派生类私有数据成员
};
void CGraduateStudent::display_GraduateStudent ()      //定义派生类公用成员函数
{
    display();          //引用基类的公用成员函数，进一步访问基类的保护成员，合法
    cout<<"sex: "<<sex<<endl;                           //引用基类的保护成员，合法
    cout<<"age: "<<age<<endl;                           //引用派生类的私有成员，合法
    cout<<"address: "<<addr<<endl;                      //引用派生类的私有成员，合法
}
void CGraduateStudent:: set_GraduateNum (int nNo) //定义派生类公用成员函数
{
    num=nNo;                                            //引用基类的保护成员，合法
}
```

同时将 main 函数改写为：

```
int main() {
    CGraduateStudent stud1;                   //stud1是派生类CGraduateStudent类的对象
    stud1.display_GraduateStudent ();         //display_GraduateStuden()是派生类中
                                              //的公用成员函数，合法
    stud1. set_GraduateNum (10086);           //通过访问派生类公用成员函数从而访问基类
                                              //保护成员，合法
    return 0;
```

```
    }

//==================================
```

程序的运行结果为:

```
num: 0
name: noName
sex: f
age: 0
address: empty
```

保护成员和私有成员不同之处,在于把保护成员的访问范围扩展到派生类中。

在实际的编程中,公用继承方式是最常用的方式,而私有继承和保护继承方式由于对基类中的行为和属性的继承加强了限制,一般只在某些特殊场合使用。

下面介绍在不同继承方式下,基类成员在派生类中的访问属性:

```cpp
class Base
{
public:
    int m1;
protected:
    int m2;
private:
    int m3;
};
class D1 : public Base
{
    //D1 继承得到一个 public 成员 m1, 一个 protected 成员 m2
    //D1 中不能访问 Base 的私有成员 m3
};
class D2 : private Base
{
    //D2 继承得到两个 private 成员 m1 和 m2
    //D2 中不能访问 Base 的私有成员 m3
};
class D3 : protected Base
{
    //D3 继承得到两个 protected 成员 m1 和 m2
    //D3 中不能访问 Base 的私有成员 m3
};
```

通过上面学习,可以在理解的基础上记住表 5-1 的内容。

5.3.4 多级派生时的访问属性

如果有这样的派生关系:类 A 为基类,类 B 是类 A 的派生类,类 C 是类 B 的派生类,则类 C 也是类 A 的派生类。类 B 称为类 A 的**直接派生类**,类 C 称为类 A 的**间接派生类**。类 A 是类 B 的**直接基类**,是类 C 的**间接基类**。对于多级继承的情况,各个层次中的类成员的访问属性,仍然是依据表 5-1 所述继承关系的派生访问属性访问原则。例如:

```cpp
class A                              //基类
{
public:
    int i;
protected :
    void f2();
    int j;
private :
    int k;
};
class B: public A              //public方式
{
public :
    void f3();
protected :
    void f4();
private :
    int m;
};
class C: protected B           //protected方式
{
public :
    void f5();
private :
    int n;
};
```

其中，类 A 是类 B 的公用基类，类 B 是类 C 的保护基类。

各成员在不同类中的访问属性如下：

无论哪一种继承方式，在派生类中是不能访问基类的私有成员，私有成员只能被本类的成员函数所访问，毕竟派生类与基类不是同一个类。

在公用派生类中，基类的 public 成员和 protected 成员被派生类作为自己的 public 成员和 protected 成员继承下来，而基类的 private 成员已经成为不可访问的，当然接下去的间接派生类更加不可能访问基类的 private 成员。

在私有派生类中，基类的 public 成员和 protected 成员被派生类作为自己的 private 成员继承下来，而基类的 private 成员也是不可访问的，同样，间接派生类也不可能访问基类的 private 成员。

大家可以根据上述的原则，判断上面的例子分别在 A、B、C 中基类的继承成员的访问属性。

私有继承时，基类的所有 public 成员在派生类中都变成了 private。如果希望它们中的某些成员可见，在派生类中进行 public 声明即可：

```cpp
/*****************************************************************
    File name    :   f0509.cpp
    Description  :   在派生类中改变继承成员的访问属性
*****************************************************************/
#include <iostream>
```

```cpp
#include <string>
using namespace std;

class Pet
{
public:
    char eat() const
    {
        return 'a';
    }
    int speak() const
    {
        return 2;
    }
    float sleep() const
    {
        return 3.0;
    }
    float sleep(int) const
    {
        return 4.0;
    }
};

class Goldfish : Pet         //默认继承方式为私有继承
{
public:
    Pet::eat;                //公用化声明，只需要指出函数的名字，不指出参数和返回类型
    Pet::sleep;              //声明了一组重载函数
};

int main()
{
    Goldfish bob;
    cout<<bob.eat();
    cout<<bob.sleep();
    cout<<bob.sleep(1);
    //! bob.speak();//Error: private member function
}

//====================================
```

程序的运行结果为：

a34

注意：这种在派生类中声明基类成员访问属性的方法不会改变基类成员的访问属性，只是改变在派生类中对基类成员的访问属性。对于基类中的重载成员函数，派生类中的访问声明将调整基类中所有重载版本的访问属性。因此，不同访问属性的重载函数名不能使用这种访问声明语法，只能在基类中单独申明。

5.4 派生类的构造函数和析构函数

定义类对象与定义变量一样，需要对对象进行初始化。如没有为对象定义构造函数，C++会自动调用默认的构造函数。这个构造函数实际上是一个空函数，不执行任何操作。如果需要对类中的数据成员初始化，应自己定义构造函数，而构造函数的主要作用是对数据成员初始化。

在设计派生类的构造函数时，不仅要考虑派生类所增加的数据成员的初始化，还应当考虑基类的数据成员初始化。也就是说，希望在执行派生类的构造函数时，使派生类的数据成员和基类的数据成员同时得到初始化。

解决这个问题的思路是在执行派生类的构造函数时，调用基类的构造函数。

下面是一个平面坐标点类：

```cpp
class Point2d
{
public:
    //对平面坐标点的操作
protected:
    double _x, _y;
};
```

如果要使用三维坐标点，可以通过继承从这个类派生出一个新类：

```cpp
class Point3d : public Point2d
{
public:
    //对三维坐标点的操作
protected:
    double _z;
};
```

那么 Point3d 类型的对象是怎样的布局呢？派生类对象中实际上包含一个基类对象，派生类对象由其基类对象以及派生类自己的非静态数据成员构成。因此，在派生类中，使用基类的非私有成员就如同使用派生类自己的成员一样。

```cpp
/******************************************************************
    File name   : f0510.cpp
    Description : 在派生类中的构造函数
******************************************************************/
#include <iostream>
#include <string>
using namespace std;
class Point2d
{
public:
    Point2d(double x = 0.0, double y = 0.0): _x(x),_y(y)
    {
```

```cpp
        }
        double x()
        {
            return _x;
        }
        double y()
        {
            return _y;
        }
        void x(double newX)
        {
            _x = newX;
        }
        void y(double newY)
        {
            _y = newY;
        }
        void print()
        {
            cout << '(' << _x <<',' << _y<<')'<<endl;
        }
    protected:
        double _x, _y;
    };

    class Point3d : public Point2d
    {
    public:
        Point3d(double x = 0.0, double y = 0.0, double z = 0.0): Point2d(x, y), _z(z)
        {}
        double z()
        {
            return _z;
        }
        void z(double newZ)
        {
            _z = newZ;
        }
        void print()
        {
            cout << '(' << x() <<',' << y() <<',' << z() << ')' <<endl;
                        //调用基类的公用成员函数x(),y()访问基类_x,_y
        }
    protected:
        double _z;
    };
        void main()
        {
```

```
        Point2d  p1(1,2);
        Point3d  p2(3,4,5);
        p1.print();
        p2.print();
    }

    //==================================
```

程序的运行结果为：

```
(1,2)
(3,4,5)
```

派生类对象中包含一个基类对象，在创建派生类对象时，基类的构造函数也应该被调用，来初始化其中的基类对象。调用的次序是先调用基类构造函数，再调用派生类的构造函数。基类构造函数的调用可以是显式的，在派生类构造函数的初始化参数列表中指出：

```
    class Point2d
    {
    public:
        Point2d(double x = 0.0, double y = 0.0): _x(x), _y(y)
        {}
        //...
    protected:
        double _x, _y;
    };
    class Point3d : public Point2d
    {
    public:
        Point3d(double x = 0.0, double y = 0.0, double z = 0.0)
            : Point2d(x, y), _z(z)   //初始化参数列表指出如何调用基类构造函数,
                                     //从而初始化基类对象
        {}
    protected:
        double _z;
    };
```

在派生类的构造函数参数表中需要为基类对象提供初始值，初始化参数列表中用基类的名字来调用构造函数。这种语法和组合子对象的初始化方式相似，只不过这里在初始化参数列表中使用的是基类的名字，而后者使用成员对象的名字。

如果在派生类的构造函数中没有显式调用基类构造函数，编译器会自动调用基类的默认构造函数来初始化派生类对象中的基类子对象。

如果一个派生类中还包含其他类的对象成员，则构造函数的调用次序是：先调用基类构造函数，再调用成员的构造函数，最后调用派生类的构造函数执行。

同样的，在撤销一个派生类对象时，基类的子对象也被撤销。析构函数的调用次序和构造函数的调用次序相反，即先调用派生类的析构函数，再调用基类的析构函数。

分析一个多级派生情况下派生类的构造函数的例子：

```cpp
/**********************************************************************
    File name   :   f0511.cpp
    Description :   多级派生情况下派生类中的构造函数
**********************************************************************/
#include <iostream>
#include<string>
using namespace std;
class Student                                   //声明基类
{
public :                                        //公用部分
    Student(int n, string nam)                  //基类构造函数
    {
        cout<<"初始化基类 Student 的成员 num 和 name"<<endl;
        num=n;
        name=nam;
    }
    void display()                              //输出基类数据成员
    {
        cout<<"num:"<<num<<endl;
        cout<<"name:"<<name<<endl;
    }
    virtual ~Student()
    {
        cout<<"析构 Student 对象"<<endl;
    }
protected :                                     //保护部分
    int num;                                    //基类有两个数据成员
    string name;
};
class StudentAdd1: public Student               //声明公用派生类 StudentAdd1
{
public :
    StudentAdd1 (int n,string nam,int a):Student(n,nam)//派生类构造函数
    {
        cout<<"初始化派生类 StudentAdd1 的成员 age"<<endl;
        age=a;              //在此处只对派生类新增的数据成员初始化
    }
    void show()
    {
        display();                              //输出 num 和 name
        cout<<"age: "<<age<<endl;               //输出 age
    }
    virtual ~StudentAdd1()
    {
        cout<<"析构 StudenAdd1 对象"<<endl;
    }
private :                                       //派生类的私有数据
    int age;                                    //增加一个数据成员
```

```cpp
    };
    class StudentAdd2:public StudentAdd1      //声明间接公用派生类 StudentAdd2
    {
    public:
        StudentAdd2(int n, string nam,int a,int s): StudentAdd1 (n,nam,a)
                                                  //间接派生类构造函数
        {
            cout<<"初始化派生类 StudentAdd2 的成员 score"<<endl;
            score=s;
        }
        void show_all()                           //输出全部数据成员
        {
            show();                               //输出 num 和 name
            cout<<"score:"<<score<<endl;          //输出 age
        }
        virtual ~StudentAdd2()
        {
            cout<<"析构 StudenAdd2 对象"<<endl;
        }
    private:
        int score;                                //增加一个数据成员
    };
    void test_func()
    {
        StudentAdd2 stud(10010,"Li",17,89);
        stud.show_all();                          //输出学生的全部数据
    }
    int main()
    {
        test_func();
        return 0;
    }


    //=================================
```

程序的运行结果为:

```
初始化基类 Student 的成员 num 和 name
初始化派生类 StudentAdd1 的成员 age
初始化派生类 StudentAdd2 的成员 score
num:10010
name:Li
age: 17
score:89
析构 StudenAdd2 对象
析构 StudenAdd1 对象
析构 Student 对象
```

注意:该例子中基类和两个派生类的构造函数的写法:

基类的构造函数首部：

```
Student(int n, string nam)
```

派生类 StudentAdd1 的构造函数首部：

```
StudentAdd1 (int n, string nam],int a):Student(n,nam)
```

派生类 StudentAdd2 的构造函数首部：

```
StudentAdd2(int n, string nam,int a,int): StudentAdd1 (n,nam,a)
```

在声明 StudentAdd2 类对象时，调用 StudentAdd2 构造函数；在执行 StudentAdd2 构造函数时，先调用 StudentAdd1 构造函数；在执行 StudentAdd1 构造函数时，先调用基类 Student 构造函数。

初始化的顺序：

(1) 初始化基类的数据成员 num 和 name。
(2) 初始化 StudentAdd1 的数据成员 age。
(3) 初始化 StudentAdd2 的数据成员 score。

而析构函数的调用次序与上面构造函数的调用次序正好相反。

5.5 多重继承与虚基类

前面讨论的是单继承，即一个类是从一个基类派生而来的。常常有这样的情况：一个派生类有两个或多个基类，派生类从两个或多个基类中继承所需的属性。C++允许一个派生类同时继承多个基类，这种行为称为**多重继承**。

5.5.1 多重继承

多重继承(Multiple Inheritance，MI)是指派生类具有两个或两个以上的直接基类(direct class)。多重继承可以看做是单继承的扩展，派生类和每个基类之间的关系可以看做是一个单继承。

声明多重继承的派生类的一般形式为：

```
class 派生类名:<继承方式>基类名1 <,<继承方式>基类名2>
{
    派生类新增加的成员
};
```

有多个基类的派生类对象中会包含各个基类对象，不同的基类可以选择不同的继承方式。派生类的对象在创建时会依据基类的声明次序来调用各个基类的构造函数，构造函数的调用次序与初始化表中的排列次序无关。图 5-5 为多重继承示意图。例如：

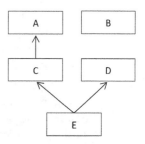

图 5-5 多重继承示意图

```
class A
{
public:
    A()
```

```cpp
        {}
    };
    class B
    {
    public:
        B()
        {}
    };
    class C : public A
    {
    public:
        C()
        {}
    };
    class D
    {
    public:
        D()
        {}
    };
    class E : public C,protected D
    {
        B member;
    public:
        E():D(),member(),C(){}
    };
```

类 E 同时继承于类 C 和类 D，属于多重继承，类 B 定义的对象 member 的是类 E 中的成员对象，逻辑上是类 E 的一个部分，这是一种**组合关系**，那么在创建一个 E 类型对象时，构造函数的调用次序是：A、C、D、B、E。析构函数的调用次序和构造函数的调用次序相反。

例如：声明一个教师(Teacher)类和一个学生(Student)类，用多重继承的方式声明一个研究生(Graduate)派生类：

教师类中包括数据成员 name(姓名)、age(年龄)、title(职称)。

学生类中包括数据成员 name1(姓名)、age(性别)、score(成绩)。

在定义派生类对象时给出初始化的数据，然后输出这些数据。

```cpp
    /*****************************************************************
        File name   : f0512.cpp
        Description : 多重继承
    *****************************************************************/
    #include <iostream>
    #include <string>
    using namespace std;
    class Teacher                                       //声明类 Teacher(教师)
    {
    public :                                            //公用部分
```

```cpp
        Teacher(string nam,int a, string t)        //构造函数
        {
            name=nam; age=a;title=t;
        }
        void display_t()                           //输出教师有关数据
        {
            cout<<"name:"<<name<<endl;
            cout<<"age"<<age<<endl;
            cout<<"title:"<<title<<endl;
        }
    protected :                                    //保护部分
        string name;
        int age;
        string title;                              //职称
};
class Student                                      //定义类Student(学生)
{
    public :
        Student(string nam,char s,float sco)       //构造函数
        {
            name=nam; sex=s; score=sco;
        }
        void display_s()                           //输出学生有关数据
        {
            cout<<"name:"<<name<<endl;
            cout<<"sex:"<<sex<<endl;
            cout<<"score:"<<score<<endl;
        }
    protected :                                    //保护部分
        string name;
        char sex;
        float score;                               //成绩
};
class Graduate:public Teacher,public Student       //声明多重继承的派生类Graduate
{
    public:
        Graduate(string nam,int a,char s, string t,float sco,float w)
            :Teacher(nam,a,t),Student(nam,s,sco),wage(w)
        {
        }
        void show()                                //输出研究生的有关数据
        {
            cout<<"name:"<<Teacher::name<<endl;
            cout<<"age:"<<age<<endl;
            cout<<"sex:"<<sex<<endl;
            cout<<"score:"<<score<<endl;
            cout<<"title:"<<title<<endl;
            cout<<"wages:"<<wage<<endl;
```

```
        }
    private:
        float wage;                                    //工资
    };
    int main()
    {
        Graduate grad1("Wang-li",24,'f',"assistant",89.5,1234.5);
        grad1.show();
        return 0;
    }

    //==================================
```

程序运行结果如下:

```
name:Wang-li
age:24
sex:f
score:89.5
title:assistant
wages:1234.5
```

在两个基类中用 name 来代表姓名,从 Graduate 类的构造函数中可以看到总参数表中的参数 nam 分别传递给两个基类的构造函数,作为基类构造函数的实参。在 show 函数调用中使用:

```
cout<<"name:"<<Teacher::name<<endl;
```

明确指出调用是哪个基类成员,这就是唯一的,不致引起二义性,能通过编译,正常运行。如果用:

```
cout<<"name:"<<name<<endl;
```

就造成继承的成员同名而产生的二义性(ambiguous)问题,导致程序无法编译运行。

在多重继承中继承的成员同名而产生的二义性问题,可以分成下列 3 种情况:

(1)两个基类有同名成员

由于基类 Teacher 和基类 Student 都有数据成员 name,编译系统无法判别要访问的是哪一基类的成员,因此程序编译出错。可以用基类名来限定 Teacher::name,从而确定唯一访问成员。

(2)两个基类和派生类三者都有同名成员

在这种情况下,派生类新增加的同名成员覆盖了基类中的同名成员,也就是说,对于派生类对象,通过其对象名访问同名的成员,则访问的是派生类的成员,而不是基类成员。特别强调的是,覆盖是指不同的成员函数,在函数名和参数个数相同、类型相匹配的情况下才发生同名覆盖。如果只有函数名相同而参数不同,不会发生同名覆盖,而属于函数重载,这样还是存在编译系统无法判别要访问的是哪一基类的成员,因此程序编译会出错。

(3)多级派生中,多个基类本身作为派生类存在共同基类

如果一个派生类有多个直接基类,而这些直接基类又有一个共同的基类,则在最终的派生类对象中会保留该间接共同基类数据成员的多份同名成员。在引用这些同名的成员时,必须在派生类对象名后增加直接基类名,以避免产生二义性,使其唯一地标识一个成员。

但是在一个类中保留间接共同基类的多份同名成员，这种现象导致二义性的问题。C++提供**虚基类**（virtual base class）的方法，使得在继承间接共同基类时只保留一份成员，从而避免了上面现象的发生。

5.5.2 虚基类

如果一个派生类有多个直接基类，而这些直接基类又有一个共同的基类，则在最终的派生类中会保留该间接共同基类数据成员的多份同名成员。在引用这些同名的成员时，必须在派生类对象名后增加直接基类名，以避免产生二义性，使其唯一地标识一个成员。

C++提供了另一种语义的继承机制：**虚继承**。在虚继承下，无论基类在派生层次中出现多少次，在派生类对象中只有一个共享的基类子对象。这样的基类被称为**虚基类**。在虚继承下，由于基类子对象的重复出现而引起的二义性就被消除了。

用 virtual 关键字修改一个基类的声明可以将它指定为虚拟基类。

```
class A {};
class B : public virtual A {};
class C : public virtual A {};
class D : public B, public C {};
```

这样 A 类的子对象在 D 中只出现一次。那么这个子对象是什么时候创建的呢？虚基类的构造函数是由最终产生对象的那个派生类的构造函数来调用的。

无论虚基类出现在继承层次中的哪个位置上，它们都是在非虚基类之前被构造。有虚基类的派生类的构造函数调用次序是：编译器按照直接基类的声明次序，检查虚基类的出现情况，每个继承子树按照深度优先的顺序检查并调用虚拟基类的构造函数。虚基类的多次出现只调用一次构造函数。虚基类的构造函数调用之后，再按照声明的顺序调用非虚基类的构造函数。例如，下面的两个程序的类的继承关系，如图 5-6 和图 5-7 所示。

第 1 个：

```
class A {};
class B {};
class C: public B, public virtual A {};
class D: public virtual A {};
class E: public C, public virtual D {};
```

E 类对象的构造函数的调用次序是：A、D、B、C、E。

第 2 个：

```
class A {};
class B : public A {};
class C {};
class D {};
class E: public virtual D {};
class F: public B, public E, public virtual C {};
```

F 类对象的构造函数的调用次序是：D、C、A、B、E、F。

下例是在例 f0512.cpp 的基础上，在 Teacher 类和 Student 类之上增加一个共同的基类 Person。将人员的一些基本数据都放在 Person 中，在 Teacher 类和 Student 类中再增加一些必要的数据。

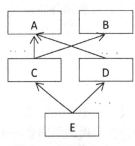
图 5-6 虚继承(一)

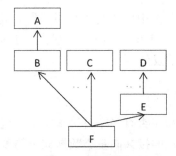
图 5-7 虚继承(二)

```
/******************************************************************
    File name  :  f0513.cpp
    Description :  虚基类
******************************************************************/
#include <iostream>
#include <string>
using namespace std;
class Person                              //声明公共基类 Person
{
public:
    Person(string nam,char s,int a)       //构造函数
    {
        name=nam;sex=s;age=a;
    }
protected :                               //保护成员
    string name;
    char sex;
    int age;
};
class Teacher:virtual public Person       //声明 Person 为公用继承的虚基类
{
public :                                  //公用部分
    Teacher(string nam,char s,int a, string t):Person(nam,s,a)
                                          //构造函数
    {
        title=t;
    }
protected :                               //保护部分
    string title;                         //职称
};
class Student:virtual public Person       //声明 Person 为公用继承的虚基类
{
public:
    Student(string nam,char s,int a,float sco) :Person(nam,s,a),score(sco)
                                          //构造函数
    {}
protected :                               //保护成员
    float score;                          //成绩
```

```cpp
    };

    class Graduate:public Teacher,public Student  //Teacher和Student为直接基类
    {
    public:
        Graduate(string nam,char s,int a, string t,float sco,float w)
                                                //构造函数
            :Person(nam,s,a),Teacher(nam,s,a,t),Student(nam,s,a,sco),wage(w)
                                                //初始化表
        {}
        void show()                              //输出研究生的有关数据
        {
            cout<<"name:"<<name<<endl;
            cout<<"age:"<<age<<endl;
            cout<<"sex:"<<sex<<endl;
            cout<<"score:"<<score<<endl;
            cout<<"title:"<<title<<endl;
            cout<<"wages:"<<wage<<endl;
        }
    private:
        float wage;                              //工资
    };
    int main()
    {
        Graduate grad1("Wang-li",'f',26,"assistant",90,3055.5);
        grad1.show();
        return 0;
    }

    //====================================
```

程序运行结果如下：

```
name:Wang-li
age:26
sex:f
score:90
title:assistant
wages:3055.5
```

因为多重继承的二义性问题，在设计中对多重继承的使用存在争议。我们主张能用单一继承解决的问题就尽量不要使用多重继承。

5.6 基类与派生类的转换

继承最重要的特性之一是公用派生类是基类真正的子类型，它完整地继承了基类的功能。这意味着公用派生类中包含从基类继承的成员，因此基类的指针和引用可以指向派生类的对象。

```cpp
/******************************************************************
    File name    : f0514.cpp
    Description : 基类与派生类的转换
******************************************************************/
#include <iostream>
#include <string>
using namespace std;
class Point2d
{
public:
    Point2d(double x = 0.0, double y = 0.0): _x(x), _y(y)
    {
    }
    double x()
    {
        return _x;
    }
    double y()
    {
        return _y;
    }
    void x(double newX)
    {
        _x = newX;
    }
    void y(double newY)
    {
        _y = newY;
    }
    void print()
    {
        cout << '(' << x() <<',' << y() << ')';
    }
protected:
    double _x, _y;
};
class Point3d : public Point2d
{
public:
    Point3d(double x = 0.0, double y = 0.0, double z = 0.0)
        : Point2d(x, y), _z(z)
    {}
    double z()
    {
        return _z;
    }
    void z(double newZ)
    {
```

```cpp
            _z = newZ;
        }
        void print()
        {
            cout << '(' << x() <<',' << y() <<',' << z() << ')';
        }
    protected:
        double _z;
};
int main()
{
    Point2d p2(1,2), *pt2 = &p2;
    Point3d p3(4, 5, 6), *pt3 = &p3;
    pt2 -> print();                    //Point2d::print()
    pt3 -> print();                    //Point3d::print()
    p2 = p3;                           //对:向上类型转换——对象切片
    p2.print();                        //Point2d::print()
    Point2d& r2 = p3;                  //对:向上类型转换
    r2.print();                        //Point2d::print()
    pt2 = &p3;                         //对:向上类型转换
    pt2 -> print();                    //Point2d::print()
}
//====================================
```

程序运行结果如下：

(1,2)(4,5,6)(4,5)(4,5)(4,5)

上面的例子中，出现了从派生类到基类的类型转换。因为在继承层次中基类位于派生类上层，这种转换被称为派生类到基类的向上转换。除了赋值操作，在函数调用期间也可能发生**向上类型转换**。C++允许这样从更特殊的类型到一般的类型的转换。

派生类对象在向基类对象转换时，会选择剩下适合的基类子对象，这就是**对象切片**。可以理解为基类子对象当它复制到一个新对象时，去掉原来对象的一部分，而不是像使用指针或引用那样简单地改变地址的内容。

指针和引用在进行转换时虽然不会发生去掉原来对象的一部分问题，但同样会**损失对象的类型信息**。因此，在上面的例子中，虽然 r2 是 Point3d 对象的引用，pt2 指向了 Point3d 类型的对象，可是通过它们调用成员函数时，调用的都是 Point2d 的成员函数。因为，编译器只知道 r2 和 pt2 是 Point2d 类型的，并不知道它们实际指向的是 Point3d 类型的对象。如果希望根据指针实际指向对象的类型来实施成员函数的调用，这就是下一章要讲的面向对象的另一大特性**多态性**。

5.7 综合应用例题

先建立一个 Point(点)类，包含数据成员 x,y(坐标点)。以它为基类，派生出一个 Circle(圆)类，增加数据成员 r(半径)，再以 Circle 类为直接基类，派生出一个 Cylinder(圆柱体)类，再增加数据成员 h(高)。并且重载运算符"<<"和">>",使之能用于输出以上类对象。

在对象化编程中,先确定抽象数据类型,然后根据确定的抽象数据类型进行抽象编程。对于简单的问题,一般可以执行下面的步骤:

(1) 找出类。
(2) 描述类与类之间的关系。
(3) 用类来界定抽象层次,从而组织程序结构。

问题描述中类的种类和关系已经给出,因此根据题意先声明基类,再声明派生类,逐级进行,分步调试。

```cpp
/****************************************************************
    File name   : f0515.cpp
    Description : 综合应用实例
****************************************************************/
//(1) 声明基类 Point 类
//可写出声明基类 Point 的部分如下:
#include <iostream>
using namespace std;
class Point                                     //声明基类 Point
{
public:
    Point(float x=0,float y=0);                 //有默认参数的构造函数
    void setPoint(float,float);                 //设置坐标值
    float getX() const                          //读 x 坐标
    {
        return x;
    }
    float getY() const                          //读 y 坐标
    {
        return y;
    }
    friend ostream & operator<<(ostream &,const Point &);//重载运算符"<<"
protected:                                      //受保护成员
    float x,y;
};

Point::Point(float a,float b)                   //Point 的构造函数,对 x,y 初始化
{
    x=a;y=b;
}
void Point::setPoint(float a,float b)           //设置 x 和 y 的坐标值
{
    x=a;y=b;
}
ostream & operator<<(ostream &output,const Point &p)    //重载运算符"<<"
{
    output<<"["<<p.x<<","<<p.y<<"]"<<endl;
    return output;
```

```cpp
}
//(2) 声明派生类 Circle
//在上面的基础上,再写出声明派生类 Circle 的部分:
class Circle:public Point                    //Point 类的公用派生类 Circle 的声明
{
public:
    Circle(float x=0,float y=0,float r=0);          //构造函数
    void setRadius(float);                          //设置半径值
    float getRadius() const;                        //读取半径值
    float area () const;                            //计算圆面积
    friend ostream &operator<<(ostream &,const Circle &);
                                                    //重载运算符"<<"
protected:
    float radius;
};

Circle::Circle(float a,float b,float r):Point(a,b),radius(r)
                                                //定义派生类 Circle 的构造函数
{}
void Circle::setRadius(float r)          //设置半径值
{
    radius=r;
}
float Circle::getRadius() const          //读取半径值
{
    return radius;
}
float Circle::area() const               //计算圆面积
{
    return 3.14159*radius*radius;
}
ostream &operator<<(ostream &output,const Circle &c)    //重载运算符"<<"
{
    output<<"Center=["<<c.x<<","<<c.y<<"],r="<<c.radius<<",area="
            <<c.area()<<endl;
    return output;
}

//(3) 声明 Circle 的派生类 Cylinder
//前面已从基类 Point 派生出 Circle 类,现在再从 Circle 派生出 Cylinder 类。
class Cylinder:public Circle            //Circle 的公用派生类 Cylinder 的声明
{
public:
    Cylinder (float x=0,float y=0,float r=0,float h=0);   //构造函数
    void setHeight(float);                                //设置圆柱高
    float getHeight() const;                              //读取圆柱高
    float area() const;                                   //计算圆表面积
```

```cpp
        float volume() const;                                   //计算圆柱体积
        friend ostream& operator<<(ostream&,const Cylinder&);//重载运算符"<<"
protected:
        float height;                                           //圆柱高
};

Cylinder::Cylinder(float a,float b,float r,float h)
    :Circle(a,b,r),height(h)                                    //定义构造函数
{
}
void Cylinder::setHeight(float h)                               //设置圆柱高
{
    height=h;
}
float Cylinder::getHeight() const                               //读取圆柱高
{
    return height;
}
float Cylinder::area() const                                    //计算圆表面积
{
    return 2*Circle::area()+2*3.14159*radius*height;
}
float Cylinder::volume() const                                  //计算圆柱体积
{
    return Circle::area()*height;
}
ostream &operator<<(ostream &output,const Cylinder& cy)         //重载运算符"<<"
{
    output<<"Center=["<<cy.x<<","<<cy.y<<"],r="<<cy.radius<<",h="<<cy.height
        <<"\narea="<<cy.area()<<", volume="<<cy.volume()<<endl;
    return output;
}

//主函数:
int main()
{
    Cylinder cy1(3.5,6.4,5.2,10);        //定义 Cylinder 类对象 cy1
    cout<<"original cylinder:\nx="<<cy1.getX()<<", y="<<cy1.getY()<<", r="
        <<cy1.getRadius()<<", h="<<cy1.getHeight()<<"\narea="<<cy1.area()
        <<",volume="<<cy1.volume()<<endl;  //用系统定义的运算符"<<"输出 cy1 的数据
    cy1.setHeight(15);                   //设置圆柱高
    cy1.setRadius(7.5);                  //设置圆半径
    cy1.setPoint(5,5);                   //设置圆心坐标值 x,y
    cout<<"\nnew cylinder:\n"<<cy1;      //用重载运算符"<<"输出 cy1 的数据
    Point &pRef=cy1;                     //pRef 是 Point 类对象的引用变量
    cout<<"\npRef as a Point:"<<pRef;    //pRef 作为一个"点"输出
    Circle &cRef=cy1;                    //cRef 是 Circle 类对象的引用变量
    cout<<"\ncRef as a Circle:"<<cRef;   //cRef 作为一个"圆"输出
```

```
        return 0;
}

//======================================
```

程序运行结果如下,请读者对照程序分析。其中圆括号中的是对运行结果的文字说明,供参考。

```
original cylinder:                      (输出 cy1 的初始值)
x=3.5, y=6.4, r=5.2, h=10               (圆心坐标 x,y;半径 r,高 h)
area=496.623,volume=849.486             (圆柱表面积 area 和体积 volume)

new cylinder:                           (输出 cy1 的新值)
Center=[5,5],r=7.5,h=15                 (以[5,5]形式输出圆心坐标)
area=1060.29, volume=2650.72            (圆柱表面积 area 和体积 volume)

pRef as a Point:[5,5]                   (pRef 作为一个"点"输出)

cRef as a Circle:Center=[5,5],r=7.5,area=176.714 (cRef 作为一个"圆"输出)
```

大家可以上机测试该程序的运行。

练　习

以点类为基类,分别定义其派生类矩形类和圆类。点为直角坐标点,矩形水平放置,由左下方的顶点和长宽定义。圆由圆心和半径定义。派生类操作判断任一坐标点是在图形内,还是在图形的边缘上,还是在图形外。缺省初始化图形退化为点。要求包括复制构造函数。编程测试程序测试设计是否正确。

根据提示,完善下列程序。(注:带编号和每条下划线处,可填写一行或多行)

```cpp
#include <iostream>
#include <cmath>
using namespace std;
const double PI=3.1415926535;
class Point
{
private:
    double x,y;
public:
    Point()
    {
        x = 0; y = 0;
    }
    Point(double xv,double yv)
    {
        x = xv;y = yv;
    }
    Point(Point& pt)
```

```cpp
        (1)    //复制构造函数实现
    double getx()
    {
        return x;
    }
    double gety()
    {
        return y;
    }
    double Area()
    {
        return 0;
    }
    void Show()
    {
        cout<<"x="<<x<<' '<<"y="<<y<<endl;
    }
};

class Circle :public Point
{
        (2)    //Circle类数据成员的定义
public:
    Circle()
    {
        radius = 0;
    }
    Circle(double xv,double yv,double vv):Point(xv,yv)
    {
        radius = vv;
    }
    Circle(Circle& cc)
        (3)    //复制构造函数实现
    double Area()
    {
        return PI*radius*radius;
    }
    void Show()                          //注意怎样访问基类的数据成员
    {
        cout<<"x="<<getx()<<'\t'<<"y="<<gety()<<'\t'<<"radius="<<radius<<endl;
    }
    int position(Point &pt)
    {
        double distance = sqrt((getx()-pt.getx())*(getx()-pt.getx())
                       +(gety()-pt.gety())*(gety()-pt.gety()));
    double s=distance-radius;
        if(s==0) return 0;           //在圆上
        else if(s<0) return -1;      //在圆内
```

```
            else return 1;                //在圆外
        }
};

class Rectangle:public Point
{
    double width,length;
public:
    Rectangle()
    {
        width=0; length=0;
    }
    Rectangle(double xv,double yv,double wv,double lv):Point(xv,xv)
    {
        width = wv;
        length= lv;
    }
    Rectangle(Rectangle& rr)
      (4)    //复制构造函数实现
    double Area()
    {
        return width*length;
    }
    void Show()
    {
        cout<<"x="<<getx()<<'\t'<<"y="<<gety()<<'\t';
        cout<<"width="<<width<<'\t'<<"length="<<length<<endl;
    }
    int position(Point &pt);
};
int Rectangle::position(Point &pt)
{
    double s1,s2;
    s1 = (pt.getx()-getx()); s2=(pt.gety()-gety());
    if(((s1==0||s1==width)&&s2<=length)||((s2==0||s2==length)&&s1<=width)) return 0;
    else if(s1<width&&s2<length) return -1;         //0 在矩形上，-1 在矩形内
        else return 1;                              //1 在矩形外
}

int main()
{
    Circle cc1(3,4,5),cc2,cc3(cc1);
    Rectangle rt1(0,0,6,8),rt2,rt3(rt1);
    Point p1(0,0),p2(6,8),p3(3,3),p4(8,4),p5(8,8);
    cc1.Show();
    cc2.Show();
    rt1.Show();
    rt2.Show();
```

```cpp
cout<<"点 p1:";
p1.Show();
cout<<"矩形 rt3:"<<'\t';
rt3.Show();
switch(rt3.position(p1))
{
    case 0:cout<<"在矩形上"<<endl;break;
    case -1:cout<<"在矩形内"<<endl;break;
    case 1:cout<<"在矩形外"<<endl;break;
}
cout<<"圆 cc3:"<<'\t';
cc3.Show();
switch(cc3.position(p1)){
    case 0:cout<<"在圆上"<<endl;break;
    case -1:cout<<"在圆内"<<endl;break;
    case 1:cout<<"在圆外"<<endl;break;
}
cout<<"点 p2:";
p2.Show();
cout<<"矩形 rt3:"<<'\t';
rt3.Show();
switch(rt3.position(p2))
{
    case 0:cout<<"在矩形上"<<endl;break;
    case -1:cout<<"在矩形内"<<endl;break;
    case 1:cout<<"在矩形外"<<endl;break;
}
cout<<"圆 cc3:"<<'\t';
cc3.Show();
switch(cc3.position(p2))
{
    case 0:cout<<"在圆上"<<endl;break;
    case -1:cout<<"在圆内"<<endl;break;
    case 1:cout<<"在圆外"<<endl;break;
}
cout<<"点 p3:";
p3.Show();
cout<<"矩形 rt3:"<<'\t';
rt3.Show();
switch(rt3.position(p3))
{
    case 0:cout<<"在矩形上"<<endl;break;
    case -1:cout<<"在矩形内"<<endl;break;
    case 1:cout<<"在矩形外"<<endl;break;
}
cout<<"圆 cc3:"<<'\t';
cc3.Show();
switch(cc3.position(p3))
```

```cpp
        {
            case 0:cout<<"在圆上"<<endl;break;
            case -1:cout<<"在圆内"<<endl;break;
            case 1:cout<<"在圆外"<<endl;break;
        }
        cout<<"点 p4:";
        p4.Show();
        cout<<"矩形 rt3:"<<'\t';
        rt3.Show();
        switch(rt3.position(p4))
        {
            case 0:cout<<"在矩形上"<<endl;break;
            case -1:cout<<"在矩形内"<<endl;break;
            case 1:cout<<"在矩形外"<<endl;break;
        }
        cout<<"圆 cc3:"<<'\t';
        cc3.Show();
        switch(cc3.position(p4))
        {
            case 0:cout<<"在圆上"<<endl;break;
            case -1:cout<<"在圆内"<<endl;break;
            case 1:cout<<"在圆外"<<endl;break;
        }
        cout<<"点 p5:";
        p5.Show();
        cout<<"矩形 rt3:"<<'\t';
        rt3.Show();
        switch(rt3.position(p5))
        {
            case 0:cout<<"在矩形上"<<endl;break;
            case -1:cout<<"在矩形内"<<endl;break;
            case 1:cout<<"在矩形外"<<endl;break;
        }
        cout<<"圆 cc3:"<<'\t';
        cc3.Show();
        switch(cc3.position(p5))
        {
            case 0:cout<<"在圆上"<<endl;break;
            case -1:cout<<"在圆内"<<endl;break;
            case 1:cout<<"在圆外"<<endl;break;
        }
        return 0;
}
```

第 6 章 多态性与虚函数

多态性是面向对象程序设计语言中除了封装性和继承性之外的又一个重要特性。多态性提供了接口与具体实现之间的另一层隔离，改善了代码的结构和可读性，同时也使创建的程序具有可扩展性。C++通过虚函数和动态绑定实现多态性。

- **知识要点**

  理解面向对象技术的可扩展性；
  掌握面向对象程序设计的重要特征——多态性；
  掌握虚函数的使用方法。

- **探索思考**

  有了多态和虚函数以后，实现了"一个接口，多种方法"，这与面向对象技术的可扩展性有什么关系？

- **预备知识**

  复习类封装和继承，特别是存在继承关系的条件下，派生类对象对基类数据成员和基类成员函数的访问控制。

6.1 多态性的概念

多态性（polymorphism）是面向对象程序设计的一个重要特征。在 C++程序设计中，多态性是指具有不同功能的函数可以用同一个函数名，这样就可以用一个函数名调用不同内容的函数。

以前学过的函数重载和运算符重载也是一种多态性，不过它们是在程序编译时系统就能决定调用的是哪个函数，因此称为**静态多态性**，也称为**编译时的多态性**。还有一种多态性是**动态多态性**，是指在程序运行过程中动态地确定操作所针对的对象，因此又称为**运行时的多态性**。动态多态性是通过**虚函数**（virtual function）实现的。

通过下面的例子来理解静态多态性，又可以进一步讨论动态多态性。

上一章讲到继承描述了类型和子类型的关系——公有派生类是基类的子类型，允许派生类到基类的向上类型转换，但是这种向上类型转换会损失类型信息。

```
/****************************************************************
    File name    : f0601.cpp
    Description  : 静态绑定
****************************************************************/
#include <iostream>
using namespace std;

class Point2d
{
```

```cpp
public:
    Point2d(double x = 0.0, double y = 0.0): _x(x), _y(y)
    {}
    void print()
    {
        cout << '(' << _x <<',' << _y << ')';
    }
protected:
    double _x, _y;
};
class Point3d : public Point2d
{
public:
    Point3d(double x = 0.0, double y = 0.0, double z = 0.0): Point2d(x, y), _z(z)
    {}
    void print()
    {
        cout << '(' << _x <<',' << _y <<',' << _z << ')';
    }
protected:
    double _z;
};

void main()
{
    Point2d *pt2; Point3d p3(4, 5, 6);
    Point3d *pt3;
    pt2 = &p3;                          //向上类型转换
    pt2 -> print();                     //Point2d::print()
    pt3 = &p3;
    pt3 -> print();                     //Point3d::print()
}
//====================================
```

程序的输出结果为：

(4,5)(4,5,6)

在这个例子中 pt2 实际指向的是 Point3d 类型的对象，可是却调用了 Point2d 类的 print() 操作。为什么不能按照 pt2 实际指向的对象类型实施调用呢？这是由于该程序在编译时由编译系统已经确定应调用函数并与之关联绑定，导致 pt2 只能按照编译时已经绑定的函数进行调用，而不能按照实际指向的对象类型实施调用。在 C++中称这种函数关联绑定方式为**静态多态性的绑定方式**。

函数调用绑定：把函数体和函数调用相联系称为**绑定**(binding，也译作捆绑或编联)。C++中，默认的函数调用绑定方式是**静态绑定**，即在程序运行之前，由编译器和连接器实现。

在上面的例子中，编译器只知道 pt2 是 Point2d 的地址，所以将 print() 的调用和 Point2d 类型联系在一起，不能调用到实际对象的操作。解决这个问题的方法是使用**动态绑定**，即在程序运行时，根据对象的类型绑定函数调用。必须有某种机制来确定运行时对象类型并调用合适的成员函数。在 C++中，通过使用**虚函数**来实现**动态绑定**。

6.2 虚 函 数

6.2.1 虚函数的作用

在类的继承层次结构中，在不同的层次中可以出现名字相同、参数个数和类型都相同而功能不同的函数。编译系统按照同名覆盖的原则决定调用的对象。需要用一定的方式来区分调用的对象，在上面的 f0601.cpp 中用 pt2 -> print() 是调用了 Point2d 中的成员函数 print()。如果想调用派生类中的 Point3d 的函数 print()，需要使用 Point3d * pt3 = &p3; pt3-> print()。通过这种方法来区分两个同名的函数，是很不方便的。

人们提出这样的设想，能否用同一个调用形式，根据所指对象的实际类型，确定调用派生类或调用基类的同名函数。C++中的虚函数就是用来解决这个问题的。虚函数的作用是允许在派生类中重新定义与基类同名的函数，并且可以通过基类指针或引用来访问基类和派生类中的同名函数。

虚函数在基类中用 virtual 关键字声明。在所有派生类中，即使不再重复声明，它也是虚函数。派生类中可以重定义基类的虚函数，这称为**覆盖**或**改写**。

上面的代码可以修改为如下形式：

```
/*****************************************************************
    File name   : f0602.cpp
    Description : 动态绑定和虚函数
*****************************************************************/
#include <iostream>
using namespace std;
class Point2d
{
public:
    Point2d(double x = 0.0, double y = 0.0): _x(x), _y(y)
    {}
    virtual void print()                //虚函数
    {
        cout << '(' << _x <<',' << _y << ')';
    }
protected:
    double _x, _y;
};
class Point3d : public Point2d
{
public:
    Point3d(double x = 0.0, double y = 0.0, double z = 0.0): Point2d(x, y), _z(z)
    {}
```

```
        void print()          //也是虚函数，覆盖了Point2d::print()
        {
            cout << '(' << _x <<',' << _y <<',' << _z << ')';
        }
    protected:
        double _z;
    };
    int main()
    {
        Point2d p2(1,2), *pt2=&p2; Point3d p3(4, 5, 6);
        pt2 -> print();          //调用 Point2d::print()
        pt2 = &p3;
        pt2 -> print();          //根据实际指向的对象类型调用 Point3d::print()
    }

    //========================================
```

程序的输出结果为：

```
(1,2)(4,5,6)
```

程序运行的结果可以看出，同样是对 print()的调用，却产生了不同的效果，这就是多态性。多态性可以为程序设计带来更大的灵活性和可扩展性。

虚函数实现了在程序运行阶段，对于同一类族中不同类的对象，对同一函数调用作出不同的响应，这就是虚函数实现的动态多态性。

虚函数的使用方法如下：

(1)在基类用 virtual 声明成员函数为虚函数。

这样就可以在派生类中重新定义此函数，为它赋予新的功能，并能方便地调用。如果成员函数在类中申明为虚函数，而在类外定义该函数时不必再加 virtual。

(2)在派生类中重新定义此函数，要求函数名、函数类型、函数参数个数和类型全部与基类的虚函数相同，并根据派生类的需要重新定义函数体。

C++规定，当一个成员函数被声明为虚函数后，其派生类中的同名函数都自动成为虚函数。因此在派生类重新声明该虚函数时，可以加 virtual，也可以不加，但习惯上一般在每一层声明该函数时都加 virtual，使程序更加清晰。如果在派生类中没有对基类的虚函数重新定义，则派生类简单地继承其直接基类的虚函数。

(3)通过基类对象类型的指针变量或引用调用虚函数，此时调用的就是指针变量或引用实际指向的对象的同名函数。通过虚函数与基类对象类型的指针变量或引用的配合使用，就能方便地调用同一类族中不同类的同名函数。

有时在基类中定义的非虚函数会在派生类中被重新定义，如果用基类指针调用该成员函数，则系统会调用对象中基类部分的成员函数；如果用派生类指针调用该成员函数，则系统会调用派生类对象中的成员函数，这并不是多态性行为，没有用到虚函数的功能。

6.2.2 虚析构函数

析构函数的作用是在对象撤销之前做必要的"清理现场"的工作。当派生类的对象从内存中撤销时，一般先调用派生类的析构函数，然后再调用基类的析构函数。

但是，如果用 new 运算符建立了临时对象，若基类中有析构函数，并且定义了一个指向该基类的指针变量，在程序用带指针参数的 delete 运算符撤销对象时，会发生一种情况：系统会只执行基类的析构函数，而不执行派生类的析构函数。这就导致基类的析构函数完成了基类对象的撤销清理，但派生类对象的析构函数由于没有执行，从而导致撤销清理不彻底，可能会发生错误。由于现代的操作系统有很完善的内存管理机制，一定程度上可以避免一些类似的内存管理错误发生，但是从程序设计的角度来说，这样的程序是有缺陷的，需要更正。

下面的例子是基类中有非虚析构函数时的执行情况。

```cpp
/*******************************************************************
    File name    : f0603.cpp
    Description : 非虚析构函数
*******************************************************************/
#include <iostream>
using namespace std;
class Point                        //定义基类 Point 类
{
public:
    Point()                        //Point 类构造函数
    {}
    ~Point()                       //Point 类析构函数
    {
        cout<<"executing Point destructor"<<endl;
    }
};
class Circle:public Point          //定义派生类 Circle 类
{
public:
    Circle()                       //Circle 类构造函数
    {}
    ~Circle()                      //Circle 类析构函数
    {
        cout<<"executing Circle destructor"<<endl;
    }
private:
    int radius;
};
int main()
{
    Point *p=new Circle;           //用 new 开辟动态存储空间
    delete p;                      //用 delete 释放动态存储空间
    return 0;
}
//===================================
```

程序的输出结果为：

```
executing Point destructor
```

这个示意程序中，p 是指向基类的指针变量，指向 new 开辟的动态存储空间，希望用 detele 释放 p 所指向的空间。

运行结果表明程序只执行了基类 Point 的析构函数，而没有执行派生类 Circle 的析构函数。

由于基类中的析构函数是非虚析构函数，如果希望能执行派生类 Circle 的析构函数，可以将基类的析构函数声明为虚析构函数。例如：

```cpp
/*****************************************************************
    File name    : f0604.cpp
    Description  : 虚析构函数
*****************************************************************/
#include <iostream>
using namespace std;
class Point                         //定义基类 Point 类
{
public:
    Point()                         //Point 类构造函数
    {}
    virtual ~Point()                //Point 类析构函数
    {
        cout<<"executing Point destructor"<<endl;
    }
};

class Circle:public Point           //定义派生类 Circle 类
{
public:
    Circle()                        //Circle 类构造函数
    {
    }
    ~Circle()                       //Circle 类析构函数
    {
        cout<<"executing Circle destructor"<<endl;
    }
private:
    int radius;
};
int main()
{
    Point *p=new Circle;            //用 new 开辟动态存储空间
    delete p;                       //用 delete 释放动态存储空间
    return 0;
}

//======================================
```

程序的输出结果为：

```
executing Circle destructor
executing Point destructor
```

运行的过程，先调用了派生类的析构函数，再调用了基类的析构函数。

当基类的析构函数定义为虚函数时，无论指针指的是同一类族中的哪一个类对象，系统会采用动态绑定，调用相应的析构函数，对该对象进行清理工作。

如果将基类的析构函数声明为虚函数时，由该基类所派生的所有派生类的析构函数也都自动成为虚函数，即使派生类的析构函数与基类的析构函数名字不相同。

因此，编程时通常把基类的析构函数声明为虚函数。

6.3 纯虚函数与抽象类

6.3.1 纯虚函数

要实现虚函数的多态调用，必须通过基类的指针或引用。这意味着在设计时，为了给一组派生类提供一个公共接口，需要设计一个没有全部实现或者全部都不实现的基类。例如，设计一个基类 Shape，并在其中定义 area() 成员，基类 Shape 仅仅作为其派生类的接口，也就是只想用基类进行向上类型转换，使用它的接口，而并不希望实际创建基类的对象，这样在基类中 area() 就可以声明成一个纯虚函数 area()，然后让 Point 类继承此类。

```
class Shape
{
public:
    virtual double area() = 0;
    //其他成员的声明…
};
```

纯虚函数的定义形式为：

```
virtual 函数类型    函数名(参数表列)=0;
```

注意：
(1) 纯虚函数没有函数体。
(2) 最后面的"=0"并不表示函数返回值为 0，它只起形式表示该函数是纯虚函数。
(3) 这是一个声明语句，最后应有分号。纯虚函数只有函数的名字而不具备函数的功能，不能被调用。只有在派生类中对此函数提供定义后，才能具备函数的功能，这时才可被调用。纯虚函数的作用是在基类中为其派生类保留一个函数的名字，以便派生类根据需要对其进行定义。

如果在一个类中声明了纯虚函数，而在其派生类中没有对该函数定义，则该虚函数在派生类中仍然为纯虚函数。

有了纯虚函数，要让它发挥作用，必须要理解另外一个概念**抽象类**。关于抽象类将在下一节讨论。

6.3.2 抽象类

创建一个纯虚函数允许在类的接口中出现没有实现或不需要实现的操作。包含至少一个纯虚函数的类是抽象类(abstract class)。如果一个抽象类中的所有成员函数都是纯虚函数，这个类称为纯抽象类。

抽象类的使用有一些限制。不能创建抽象类的实例，但是可以声明抽象类的指针或引用。事实上，我们也是通过抽象基类的指针或引用来实现虚函数的多态调用。

当继承一个抽象类时，必须在派生类中实现所有的纯虚函数，否则派生类也被看做是一个抽象类。

抽象类的作用是作为一个类族的共同基类，或者说，它提供一个公共接口。抽象类不能实例化，但是可以定义指向抽象类数据的指针变量。当派生类成为具体类之后，就可以用这种指针指向派生类对象，然后通过该指针调用虚函数，实现多态性的操作。

6.3.3 抽象类应用实例

5.7 节综合应用实例中介绍了以 Point 为基类的点—圆—圆柱体类的层次结构。现在要对其进行改写，在程序中使用虚函数和抽象基类。类的层次结构的顶层是抽象基类 Shape（形状）。Point（点）、Circle（圆）、Cylinder（圆柱体）都是 Shape 类的直接派生类和间接派生类。

下面是一个完整的程序，为了便于阅读，分段插入了一些文字说明。

```cpp
/******************************************************************
    File name   : f0605.cpp
    Description : 纯虚函数和抽象类应用实例
******************************************************************/
//第1部分
#include <iostream>
using namespace std;
class Shape                                    //声明抽象基类 Shape
{
public:
    virtual double area() const =0;            //纯虚函数
    virtual double volume()const =0;           //纯虚函数
    virtual void shapeName()const =0;          //纯虚函数
};

//第2部分
class Point:public Shape                       //声明 Shape 的公用派生类 Point
{
public:
    Point(double=0,double=0);
    void setPoint(double,double);
    double getX()const
    {
        return x;
    }
    double getY()const
    {
        return y;
    }
    virtual double area() const
    {
        return 0.0;
```

```cpp
    };
    virtual double volume()const
    {
        return 0.0;
    }
    virtual void shapeName()const              //对虚函数进行再定义
    {
        cout<<"Point:\n";
    }
    friend ostream & operator <<(ostream &,const Point &);
protected:
    double x,y;
};

//定义 Point 类成员函数
Point::Point(double a,double b)
{
    x=a;
    y=b;
}
void Point::setPoint(double a,double b)
{
    x=a;
    y=b;
}
ostream & operator <<(ostream &output,const Point &p)
{
    output<<"["<<p.x<<","<<p.y<<"]";
    return output;
}

//第 3 部分
class Circle:public Point                      //声明 Circle 类
{
public:
    Circle(double x=0,double y=0,double r=0);
    void setRadius(double);
    double getRadius()const;
    virtual double area() const;
    virtual double volume()const
    {
        return 0.0;
    }
    virtual void shapeName()const              //对虚函数进行再定义
    {
        cout<<"Circle:\n";
    }
    friend ostream &operator <<(ostream &,const Circle &);
```

```cpp
protected :
    double radius;
};

//声明 Circle 类成员函数
Circle::Circle(double a,double b,double r):Point(a,b),radius(r)
{
}
void Circle::setRadius(double r)
{
    radius=r;
}
double Circle::getRadius()const
{
    return radius;
}

double Circle::area()const
{
    return 3.14159*radius*radius;
}
ostream &operator <<(ostream &output,const Circle&c)
{
    output<<"["<<c.x<<","<<c.y<<"], r="<<c.radius;
    return output;
}


//第 4 部分
class Cylinder:public Circle                    //声明 Cylinder 类
{
public:
    Cylinder(double x=0,double y=0,double r=0,double h=0);
    void setHeight(double);
    virtual double area()const;
    virtual double volume()const;
    virtual void shapeName()const               //对虚函数进行再定义
    {
        cout<<"Cylinder:\n";
    }
    friend ostream& operator<<(ostream&,const Cylinder&);
protected:
    double height;
};

//定义 Cylinder 类成员函数
Cylinder::Cylinder(double a,double b,double r,double h):Circle(a,b,r),height(h)
{
}
```

```cpp
void Cylinder::setHeight(double h)
{
    height=h;
}
double Cylinder::area()const
{
    return  2*Circle::area()+2*3.14159*radius*height;
}
double Cylinder::volume()const
{
    return  Circle::area()*height;
}

ostream &operator <<(ostream &output,const Cylinder& cy)
{
    output<<"["<<cy.x<<","<<cy.y<<"], r="<<cy.radius<<", h="<<cy.height;
    return  output;
}

//第5部分
int main()                                              //main 函数
{
    Point point(3.2,4.5);                               //建立 Point 类对象 point
    Circle circle(2.4,1.2,5.6);                         //建立 Circle 类对象 circle
    Cylinder cylinder(3.5,6.4,5.2,10.5);                //建立 Cylinder 类对象
    point.shapeName();                                  //静态绑定
    cout<<"\t"<<point<<endl;
    circle.shapeName();                                 //静态绑定
    cout<<"\t"<<circle<<endl;
    cylinder.shapeName();                               //静态绑定
    cout<<"\t"<<cylinder<<endl;
    Shape *pt;                                          //定义基类指针
    pt=&point;                                          //指针指向 Point 类对象
    pt->shapeName();                                    //动态绑定
    cout<<"\tx="<<point.getX()<<",y="<<point.getY()<<"\n\tarea="<<pt->area()
        <<"\n\tvolume="<<pt->volume()<<"\n";
    pt=&circle;                                         //指针指向 Circle 类对象
    pt->shapeName();                                    //动态绑定
    cout<<"\tx="<<circle.getX()<<",y="<<circle.getY()<<"\n\tarea="<<pt->area()
        <<"\n\tvolume="<<pt->volume()<<"\n";
    pt=&cylinder;                                       //指针指向 Cylinder 类对象
    pt->shapeName();                                    //动态绑定
    cout<<"\tx="<<cylinder.getX()<<",y="<<cylinder.getY()<<"\n\tarea="<<pt->area()
        <<"\n\tvolume="<<pt->volume()<<"\n";
    return 0;
}

//======================================
```

程序运行结果如下，请读者对照程序分析。（其中圆括号中的是对运行结果的文字说明供参考）

```
    Point:                      （输出Point类对象point的数据：点的坐标）
        [3.2,4.5]
    Circle:                     （输出Circle类对象circle的数据：圆心和半径）
        [2.4,1.2], r=5.6
    Cylinder:                   （输出Cylinder类对象cylinder的数据：圆心、半径和高）
        [3.5,6.4], r=5.2, h=10.5
    Point:
        x=3.2,y=4.5             （输出Point类对象point的数据：点的坐标）
        area=0                  （输出Point类对象point的数据：点的面积）
        volume=0                （输出Point类对象point的数据：点的体积）
    Circle:
        x=2.4,y=1.2             （输出Circle类对象circle的数据：圆心坐标）
        area=98.5203            （输出Circle类对象circle的数据：圆的面积）
        volume=0                （输出Circle类对象circle的数据：圆的体积）
    Cylinder:
        x=3.5,y=6.4             （输出Cylinder类对象cylinder的数据：圆心坐标）
        area=512.959            （输出Cylinder类对象cylinder的数据：圆柱的表面积）
        volume=891.96           （输出Cylinder类对象cylinder的数据：圆柱的体积）
```

从本例可以进一步明确以下结论：

虚函数被应用于多态调用，多态调用需要通过基类的指针或引用。而一般作为基类抽象类给一组派生类提供一个公共的接口，从而把类的声明与类的使用分离。这对于设计类库的软件开发者来说尤为重要。软件开发者设计了各种各样的类，不用向用户提供源代码，用户可以不知道类是怎样声明的，但是可以使用这些类来派生出自己的类。实现了"一个接口，多种方法"，面向对象技术的可扩展性就体现出来了。

6.4 综合应用例题

在银行账户中，一个基础账号 Account 类，包含账号和余额信息，作为基类。一个从基础账号 Account 类派生的存款账号 Savings 类，包含最小余额信息。另一个从基础账号 Account 类派生的结算账号类 Checking 类，包含汇款方式信息。

AccountList 类定义了一个链表，通过链表可以将银行的各个账户对象组织在这个双向链表中，从而实现了类似容器的功能。在链表中通过指针来操作元素，可以体现出元素的多态性。

程序采用多文件结构，包含主程序文件在内一共 9 个文件，分别是：

基础账号类：

(1) account.h

(2) account.cpp

存款账号类：

(3) savings.h

(4) savings.cpp

结算账号类：

(5) checking.h
(6) checking.cpp

链表容器类：
(7) accountlist.h
(8) accountlist.cpp

主程序：
(9) linked_list.cpp

基础账号 Account 类因为包含纯虚函数 withdrawal(double)——取款，所以是抽象类，在其派生类 Savings 和 Checking 中分别实现了该虚函数。另外，还包含了普通的虚函数 display()；

```
/******************************************************************
    File name    : account.h
    Description : 综合应用实例 account.h 部分
******************************************************************/
#ifndef HEADER_ACCOUNT
#define HEADER_ACCOUNT
//-----------------------------------
#include<string>
using std::string;
//-----------------------------------
class Account
{
protected:
    string acntNumber;                          //账号
    double balance;                             //余额
public:
    Account(string acntNo, double balan=0.0);
    virtual void display()const;
    double getBalan()const
    {
        return balance;
    }
    void deposit(double amount)
    {
        balance += amount;
    }
    bool operator==(const Account& a)
    {
        return acntNumber==a.acntNumber;
    }
    string getAcntNo()
    {
        return acntNumber;
    }
    virtual void withdrawal(double amount)=0;
};
//-----------------------------------
```

```cpp
#endif                                          //HEADER_ACCOUNT

/************************************************************************
    File name    : account.cpp
    Description : 综合应用实例 account.cpp 部分
************************************************************************/
#include<iostream>
#include"account.h"
//-------------------------------------
Account::Account(string acntNo, double balan)
    :acntNumber(acntNo),balance(balan)
{}
//-------------------------------------

void Account::display()const
{
    std::cout<<"Account:"+acntNumber+" = "<<balance<<"\n";
}
//-------------------------------------


/************************************************************************
    File name    : savings.h
    Description : 综合应用实例 savings.h 部分
************************************************************************/
#ifndef HEADER_SAVINGS
#define HEADER_SAVINGS
//-------------------------------------
#include"account.h"
#include<string>
using std::string;
//-------------------------------------
class Savings: public Account
{
public:
    Savings(string acntNo, double balan=0.0):Account(acntNo,balan)
    {}
    void display()const;
    void withdrawal(double amount);
};
//-------------------------------------
#endif                                          //HEADER_SAVINGS

/************************************************************************
    File name    : savings.cpp
    Description : 综合应用实例 savings.cpp 部分
************************************************************************/
#include<iostream>
#include"savings.h"
```

```cpp
//-----------------------------------
void Savings::display()const
{
    std::cout<<"Savings ";
    Account::display();
}
//-----------------------------------
void Savings::withdrawal(double amount)
{
    if(balance < amount)
        std::cout <<"Insufficient funds withdrawal: "<<amount<<"\n";
    else
        balance -= amount;
}
//-----------------------------------


/******************************************************************
    File name    : checking.h
    Description : 综合应用实例 checking.h 部分
******************************************************************/
#ifndef HEADER_CHECKING
#define HEADER_CHECKING
#include"account.h"
//-----------------------------------
enum REMIT{remitByPost, remitByCable, other};    //信汇、电汇、无
//-----------------------------------
class Checking: public Account
{
    REMIT remittance;                           //汇款方式
public:
    Checking(string acntNo, double balan=0.0);
    void withdrawal(double amount);
    void display()const;
    void setRemit(REMIT re)
    {
        remittance = re;
    }
};
//-----------------------------------
#endif                                          //HEADER_CHECKING


/******************************************************************
    File name    : checking.cpp
    Description : 综合应用实例 checking.cpp 部分
******************************************************************/
#include<iostream>
#include"checking.h"
```

```cpp
Checking::Checking(string acntNo, double balan)
    :Account(acntNo, balan), remittance(other)
{
}
//------------------------------------
void Checking::display()const
{
    std::cout<<"Checking ";
    Account::display();
}
//------------------------------------
void Checking::withdrawal(double amount)
{
    if(remittance==remitByPost)                 //信汇30元手续费
        amount += 30;
    if(remittance==remitByCable)                //电汇60元手续费
        amount += 60;
    if(balance < amount)
        std::cout <<"Insufficient funds withdrawal: "<<amount<<"\n";
    else
        balance -= amount;
}

/******************************************************************
    File name    : accountlist.h
    Description : 综合应用实例 accountlist.h 部分
******************************************************************/
#ifndef ACCOUNTLIST
#define ACCOUNTLIST
#include"account.h"
//------------------------------------
class Node
{
public:
    Account& acnt;
    Node *next, *prev;                          //双向链表,每个节点前后指针
    Node(Account& a):acnt(a),next(0),prev(0)
    {}
    bool operator==(const Node& n)const
    {
        return acnt==n.acnt;
    }
};
//------------------------------------

class AccountList
{
    int size;
```

```cpp
        Node *first;
    public:
        AccountList():first(0),size(0)
        {}
        Node* getFirst()const
        {
            return first;
        }
        int getSize()const
        {
            return size;
        }
        void add(Account& a);
        void remove(string acntNo);
        Account* find(string acntNo)const;
        bool isEmpty()const
        {
            return !size;
        }
        void display()const;
        ~AccountList();
    };
    #endif                                            //HEADER_ACCOUNTLIST

    /**********************************************************************
        File name   : accountlist.cpp
        Description : 综合应用实例 accountlist.cpp 部分
    **********************************************************************/
    #include<iostream>
    using std::cout;
    #include"accountlist.h"
    //------------------------------------
    void AccountList::add(Account& a)
    {
        Node* pN = new Node(a);
        if(first)
        {
            pN->next = first;
            first->prev = pN;
        }
        first = pN;
    size++;
    }
    //------------------------------------
    void AccountList::remove(string acntNo)
    {
        for(Node* p=first; p; p=p->next)
        if((p->acnt).getAcntNo()==acntNo)
```

```cpp
        {
            if(p->prev) p->prev->next = p->next;
            if(p->next) p->next->prev = p->prev;
            if(p==first) first = p->next;
            delete p;
            size--;
            break;
        }
}
//------------------------------------
Account* AccountList::find(string acntNo)const
{
    for(Node* p=first; p; p=p->next)
        if((p->acnt).getAcntNo()==acntNo)
            return &(p->acnt);
    return 0;
}
//------------------------------------
void AccountList::display()const
{
    std::cout<<"There are "<<size<<" accounts.\n";
        for(Node* p=first; p; p=p->next)
            (p->acnt).display();
}
//------------------------------------
AccountList::~AccountList()
{
    for(Node* p=first; p=first; delete p)
        first = first->next;
}
//------------------------------------


/************************************************************************
File name   : linked_list.cpp
Description : 综合应用实例主程序部分
************************************************************************/
#include<iostream>
using namespace std;
#include"savings.h"
#include"checking.h"
#include"accountlist.h"
int main()
{
    Savings s1("101",2000), s2("102", 4000);
    Checking c1("201"), c2("312", 10000);
    AccountList a;
    a.add(s1);
    a.add(s2);
```

```
        a.add(c1);
        a.add(c2);
        Account* p;
        if(p = a.find("101")) p->deposit(100);
        if(p = a.find("201")) p->deposit(2000);
        if(p = a.find("102")) p->withdrawal(2500);
        if(p = a.find("312")) p->withdrawal(2950);
        a.display();
}

//===================================
```

程序的运行结果:

```
There are 4 accounts.
Checking Account:312 = 7050
Checking Account:201 = 2000
Savings Account:102 = 1500
Savings Account:101 = 2100
```

在将不同的对象放到容器之后,可以进行查询对象的操作,用 virtual 定义的 withdrawal 函数和 display 函数,显现出多态性。对象进入容器后,无须用户分辨,可以自动对不同的账户采取不同的取款和显示操作。

练 习

定义一个 Shape 抽象类,派生出 Rectangle 类和 Circle 类,计算各派生类对象的面积 Area()。根据提示,完善下列程序。(注:带编号和每条下划线处,可填写一行或多行)

```
#include<iostream>
#include<string>
using namespace std;
class Shape
{
public:
    (1)    //纯虚函数 Area 的定义
};
class Rectangle:public Shape
{
public:
    Rectangle(double w,double h)
    (2)    //Rectangle 构造函数初始化成员
    double Area()
    {
        (3)    //Rectangle 中 Area 的实现
    }
protected:
    double width,height;
```

```
};

class Circle:public Shape
{
public:
    Circle(double r)
         (4)    //Circle 构造函数初始化成员
    double Area()
    {
         (5)   //Circle 中 Area 的实现
    }
protected:
    double radius;
};

int main()
{
    Shape *sp;
    Rectangle re1(10,6);
    Circle cir1(4.0);
    sp=&re1;
    sp->Area ();
    sp=&cir1;
    sp->Area ();
    return 0;
}
```

第7章 文件系统

文件的读写在面向对象程序设计中具有重要的作用。在 I/O 类库的基础上，围绕文件 I/O 流，本章主要讲解了文件的基本概念、打开和关闭文件、文件读写、在文件中定位和处理成批文件。通过对本章内容的学习，可以学会建立二进制文件和文本文件，读取文件中的数据和把相应的计算结果保存到文件中。

- **知识要点**

理解有关文件的相应概念，如文本文件、二进制文件、标准文件、缓冲型文件和非缓冲型文件；掌握面向对象程序设计下的文件打开和关闭、文件读写、文件定位，特别是批文件的处理。

- **探索思考**

在 C 语言中，对文件的操作具体是怎么样进行的？
结合单一文件和批文件的具体操作过程，总结两者的异同。

- **预备知识**

回顾标准输入设备(键盘)和标准输出设备(显示器)是如何完成 C++程序数据的输入和输出的。

7.1 文件的基本概念

流文件描述了程序从何处输入数据，以及向何处输出数据。通常编写的 C++程序都是从键盘输入并向屏幕输出的。对于大批量数据来说，键盘输入不方便，而且容易出错，屏幕输出不易阅读和保存，不便用作另一个程序的输入数据，比如在要输入学生学籍信息的时候，不仅要多次输入学号，还要输入学生不同课程的成绩；在具体输入的过程中，由于学号和学习成绩比较相似，特别容易输错。

在对学生的成绩求平均值并输出计算结果的过程中，该结果不能保存并且只能阅读一次。如果用磁盘文件来存储相应的数据，问题就相对简单了，指定某个磁盘文件用作输入，指定另外某个磁盘文件用作输出，如果有需要，也可以在同一个文件中完成输入和输出的功能。

输入输出的操作指令实际上是一种数据传递的完整过程，是字符序列在内部主机和外部载体之间的交互。在前面已经通过文件的操作接触过流，流(stream)为从源(或生产者)到目的(或消费者)的数据流的引用。流具有方向性：与输入设备(如键盘)相联系的流称为输入流；与输出设备(如屏幕)相联系的流称为输出流；与输入输出设备(如磁盘)相联系的流称为输入输出流。C++中主要包含了标准输入流(cin)、标准输出流(cout)、非缓冲型标准出错流(cerr)和缓冲型的标准出错流(clog)四种类型。为了完成数据的输入和输出功能，C++自身定义了一个体系繁多的 I/O 类库，该类库涵盖了 ios、istream、ostream、iostream、ifstream、ofstream、fstream、istrstream、ostrstream、strstream 和 stdiostream 等，其中 ios 为根基类，其他则是由其直接或者间接派生出来的类，这些输入输出类之间的层次关系如图 7-1 所示。

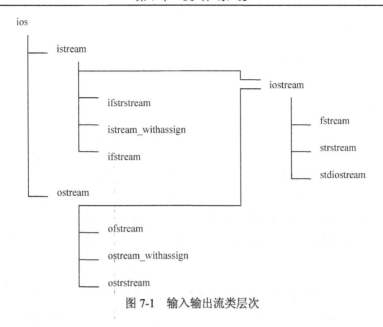

图 7-1 输入输出流类层次

7.1.1 文本文件和二进制文件

在上述介绍的有关输入和输出的基本知识基础上，为了更好地实现对文件的操作，下面介绍有关文件的概念。一般来讲，文件是以计算机磁盘为载体，存储在计算机上的信息集合。但在 C++程序开发过程中，文件不是由一条条记录构成的，而是由一个个字符(字节)序列组成的，即流式文件。根据文件中数据的呈现形式，具体分为文本文件和二进制文件：文本文件中存放的是字符，二进制文件中存放的是二进制数据，如图像、声音、视频等。数值可以用字符串方式表示，存放于文本文件中，也可以用二进制表示，存放于二进制文件中。例如，下面 10 个整数，每个整数的字符串表示都是占用 5 字节：

31444   19474   17560   17076   19460
13594   31477   13693   29516   31902

每两个整数之间有一个制表符，合计 8 个制表符，共有两行，每行有一个换行符和一个回车符，不存任何空格，文件长度为 5×10＋8＋2×2＝62 字节。若用二进制文件存放，每个整数只需 2 字节(短整数)，文件长度为 2×10＝20 字节。从这个例子看出，用文本文件存储数值会占用较大的磁盘空间，阅读和修改都很方便，当然数据的安全性也就差一些。用二进制文件存储占用空间较小，数据的安全性好，但阅读不方便，而且数据的意义依赖于程序：这 10 个短整数无法保证不被另外一个程序当作 5 个长整数读取。归根结底，字符也是一种二进制数据，因此文本文件也可以作为二进制文件来处理。对文本文件和二进制文件的输入和输出是不同的，具体实现如下。

流运算符 ">>" 表示从流中输入数据。比如，系统有一个默认的标准输入流(cin)，一般情况下是指的键盘，所以 cin>> x; 就表示从标准输入流中读取一个指定类型(即变量 x 的类型)的数据。

流运算符 "<<" 表示向流输出数据。比如，系统有一个默认的标准输出流(cout)，一般情况下是指的显示器，所以 cout<< " Write Stdout " << '\n'; 就表示把字符串" Write Stdout" 和换行字符('\n')输出到标准输出流。

二进制文件的输入和输出

```
        istream& istream::read(char *,int);         //每次从二进制流提取若干个字符
        ostream& ostream::write(const char *,int);   //每次向二进制流插入若干个字符
```

7.1.2 标准文件

与某种设备相关联的文件，称为标准文件。标准输入文件是命令的输入，缺省是键盘，也可以是文件或其他命令的输入，如 scanf、getchar、gets 等函数就是读标准输入文件。标准输出文件是命令的输出，缺省是屏幕，也可以是文件，如 printf、putchar、puts 等函数就是写标准输出文件。标准错误文件是错误命令的输出，缺省是屏幕，同样也可以是文件。当程序中有些信息不准备输出到重定向的外存文件中去，而希望直接显示在屏幕上，就可以使用标准文件作为输出目标。输入或输出重定向，就是把一般的外存文件跟设备文件关联起来。例如：

```
        D:\ textbook > copyfile < file1.txt > file2.txt
```

其中，copyfile.exe 是自己编写的一个文件复制程序，它从标准输入设备(键盘)读入一个文件，然后复制到标准输出(屏幕)上去。符号"<"表示应把 file1.txt 当作标准输入设备，符号">"表示应把 file2.txt 当作标准输出设备。于是用户既不必从键盘上输入，也看不到屏幕上的输出。

7.1.3 缓冲型文件和非缓冲型文件

如果每读写一个数据都要访问外存(典型的外存是磁盘)，读写速度会很慢，因为外存访问本来就比内存慢，若再加上频繁访问，就会不堪重负，特别是在当下的大数据环境下，更加突显出该访问方式的弊端。为了减少访问次数，可以在内存中开辟一块"缓冲区"(buffer)，通常大小为 512 字节。读文件时，一次从外存读入一批数据装满缓冲区，再从缓冲区逐个将数据赋给变量(整存零取)。写文件时，先将数据送到缓冲区，缓冲区装满后才一起送到外存(零存整取)。这样就大大加快了文件存取的速度，而且系统会自动管理这块缓冲区。

有一利必有一弊，缓冲区加快了文件访问速度，但是也留下了数据安全方面的隐患。例如，最后一批数据未充满缓冲区，关闭文件时系统会自动将这批数据写到文件中去。如果程序员忘了关闭文件，这批数据就有可能丢失。另一种文件管理方法是不要缓冲区，处理的一般是二进制文件，读写的数据往往是"大块"的。这时使用缓冲区不仅没有减少外存访问次数，还会增加缓冲区与程序数据区的数据频繁交换的时间。用这种方法处理的文件称为非缓冲型文件。一般把缓冲文件系统的输入输出称为标准输入输出(标准 I/O)，非缓冲文件系统的输入输出称为系统输入输出(系统 I/O)。

简言之，缓冲输入的流程是：键盘→输入缓冲区→程序中的输入队列，非缓冲输入的流程是：键盘→程序中的输入队列。典型的非缓冲输入函数有 getch()，缓冲输入函数 get()。通过下面的例子可以区分缓冲区对文件读写的影响。

```
        /*****************************************************************
        File name    : f0701.cpp
        Description  : 缓冲输入与非缓冲输入的不同
        *****************************************************************/

        #include <iostream>
        #include <conio.h>
        using namespace std;
```

```cpp
int main()
{
    char ch;
    ch=cin.get();
    cout<<ch;
    ch=getch();
    printf("\nYou input a '%c'\n", ch);
    while((ch=getchar())!='\n')
    {
        putchar(ch);
    }
    getchar();
    return 0;
}
//===============================
```

输入：abcd[enter]j[enter]，程序的输出结果为：

```
a
You input a 'j'
bcd
```

在上例中，在缓冲区输入类型中，执行语句 ch=cin.get() 并输入 abcd[enter]，输出只有 a。因为 cin.get() 每次只读一个字符，所以输出时只输出 ch 的值，执行一次，ch='a'。而执行非缓冲区读入语句 getch() 时，输出为：

```
You input a 'j'
```

getchar() 函数是从缓冲区直接读取字符，于是显示

```
bcd
```

输出序列'bcd'依然连续，说明输入队列不受 getch() 这一非缓冲函数的影响。

7.2 文件的打开和关闭

文件的打开就是将文件对象跟具体的外存文件联系起来，即为文件流对象和指定的磁盘文件建立关联，以便使文件流的内容流向指定的磁盘文件。文件关闭实际上就是切断该磁盘文件与文件流的关联。文件打开这一操作，具体涉及文件打开的目的、打开的方式和特殊的动作等三个方面的内容。目的主要包括读、写或者读写兼顾，方式主要是指打开的是文本文件还是二进制文件。特殊动作是指在一定条件下所包含的文件定位、数据添加与清除、文件创建与替换等。

根据文件打开方式的不同，打开文件的使用方法主要有如下六种方式构成，假如文件的名字为 file，具体见表 7-1。

表 7-1 打开文件函数的方法

编号	情况	函数调用
1	打开文本文件读出	file.open("test.txt", ios::in);
2	打开文本文件写入	file.open("test.txt", ios::out);

续表

编号	情况	函数调用
3	打开二进制文件读出	file.open("test.txt", ios::in|ios::binary);
4	打开二进制文件写入	file.open("test.txt", ios::out|ios::binary);
5	打开文本文件来读写	file.open("test.txt", ios::in|ios::out);
6	打开文本文件追加	file.open("test.txt", ios::app);

由于输入输出方式是在 ios 类库中已经界定好的，再加上绝对部分是枚举常量，所以可有多种情况供选择。具体见表 7-2。

表 7-2 文件输入输出方式功能设置

编号	输入输出方式	主要功能
1	ios::in	以输入方式打开文件
2	ios::out	以输出方式打开文件(这是默认方式)，如果已有此名字的文件，则将其原有内容全部清除
3	ios::app	以输出方式打开文件，写入的数据添加在文件末尾
4	ios::ate	打开一个已有的文件，文件指针指向文件末尾
5	ios::trunc	打开一个文件，如果文件已存在，则删除其中全部数据；如文件不存在，则建立新文件
6	ios::binary	以二进制方式打开一个文件，如不指定此方式则默认为 ASCII 方式
7	ios::nocreate	打开一个已有的文件，如文件不存在，则打开失败。nocrcate 的意思是不建立新文件
8	ios::noreplace	如果文件不存在则建立新文件，如果文件已存在则操作失败，replace 的意思是不更新原有文件
9	ios::in | ios::out	以输入和输出方式打开文件，文件可读可写
10	ios::out | ios::binary	以二进制方式打开一个输出文件
11	ios::in | ios::binar	以二进制方式打开一个输入文件

表 7-1 和表 7-2 相应说明如下：

(1) 文件对象打开函数 open 是类的函数成员，主要有如下两种语法格式：打开编程时已知数据类型的窗口对象和打开程序运行后才能确定数据类型的窗口对象。

(2) ios::in，ios::out 等是枚举常量，这些枚举常量可以用按位或运算符来组合，如 ios::in|ios::out 表示既读又写。

(3) 二进制文件也可以既读又写，也可以追加。

(4) 文件对象有三个类：fstream、ifstream 和 ofstream，如果用后两个类来定义文件对象，则默认为只读或只写，因此处理文本文件时可以只带一个文件名作为参数。例如：

```
ifstream in;
in.open("test.txt");
ofstream out;
out.open("test.txt");
```

在打开函数 open 的基础上，文件被执行相应操作后，有两种状态：要么文件打开成功，要么打开失败。失败的原因主要由以下三种情况构成：

(1) 打开来读，但所指定的文件不存在；

(2) 文件名不合法；

(3) 已经打开了太多的文件。

除了打开文件，其他操作也都可能失败，产生运行错误。如果打开失败，后续的文件操作都是无效的。因此需要检查是否打开成功，说明成功或失败时应如何处理。例如：

```
/*********************************************************************
File name   : f0702.cpp
Description : 文件的打开和显示
*********************************************************************/

#include <fstream>
#include <iostream>
using namespace std;
void main()
{
    ifstream  in;
    in.open("test.txt");
    if (! in)
    {
        cout << "Can not open file\n";
        return;
    }
    char ch;
    while((ch = in.get()) != EOF)
    {
        cout << ch;
    }
    in.close();
}
```

如果 test.txt 的内容为：

```
abc e
123 5

end
```

程序的输出结果为：

```
abc e
123 5

end
```

fstream 是使用文件对象要用到的头文件。上面示例程序的最后一个语句是关闭文件。成员函数 close() 负责将缓存中的数据释放出来并关闭文件。这个函数一旦被调用，原先的流对象就可以被用来打开其他的文件，这个文件也就可以重新被其他的进程所访问。

为防止流对象被销毁时还联系着打开的文件，ifstream 的析构函数将会自动调用关闭函数 close。关闭文件的作用是：把缓冲区剩余的内容写进输出文件；切断文件对象(或文件指针，下同)与外存文件的联系，该文件对象可以用来打开另一个外存文件。例如，程序要以同样的方式处理 3 个文件，并不需要 3 个文件对象，而是只需定义一个，然后在第 i 次循环时打开第 i 个文件，读写之后立即关闭，在第 $i+1$ 次循环时仍用这个文件对象打开第 $i+1$ 个文件，示例代码如下：

```
/*********************************************************************
File name   : f0703.cpp
```

```
Description：多个连续文件名文件的打开和按行显示
******************************************************************/

#include <sstream>
#include <fstream>
#include <iostream>
using namespace std;
void main()
{
    for(int i=1;i<=3;i++)
    {
        string b(".txt");
        string file_name;
        stringstream ss;
        ss<<i<<b;
        ss>> file_name;
        ifstream in;
        in.open(file_name.c_str());
        if (! in)
        {
            cout << "Can not open file\n";
            return;
        }
        string line;
        getline(in,line);
        while(line!="")
        {
            cout<<line<<endl;
            getline(in,line);
        }
        in.close();
    }
}
```

注意：上述代码从语句 string b(".txt")；至语句 ss>> file_name;用于生成文件名分别为 "1.txt"、"2.txt"和"3.txt"的 3 个文件，而每次均使用唯一的文件对象 in 来打开各文件。

假如文件 1.txt 的内容是：

c++　程序设计
python　程序设计

2.txt 的内容是：

布莱希特梅兰芳

莫言余华毕飞宇

3.txt 的内容是：

白日依山尽，黄河入海流

程序的输出结果为：

```
c++      程序设计
python   程序设计
布莱希特梅兰芳
白日依山尽，黄河入海流
```

当文件打开后，就可以对文件执行输入和输出的操作了，即文件读写。依据文件打开方式的不同，文本文件可以通过字符或字符串的方式进行读写，二进制文件则以数据块或格式化的方式完成读写操作。在读写次序上可以是随机的，也可以是顺序的。

7.2.1 字符级读写

字符级读写，每次读或写一个字符，用于文本文件或二进制文件。代码示例如下：

```
/*****************************************************************
File name    : f0704.cpp
Description  : 字符级读写(1)
*****************************************************************/

#include <fstream>
#include <iostream>
using namespace std;
void main()
{
    ifstream  in;
    in.open("input.txt");
    ofstream  out;
    out.open("output.txt");
    char ch;
    while(in.get(ch))
        out.put(ch);
}
```

这是一个循环控制，每次从输入文件中读一个字符，再写到输出文件中去。用成员函数 get 和 put 来读写字符。若到达文件末尾，成员函数 get 返回 0，循环结束。成员函数 get 还有不带参数的用法，如上面的核心代码可改写为：

```
/*****************************************************************
File name    : f0705.cpp
Description  : 字符级读写(2)
*****************************************************************/

#include <fstream>
#include <iostream>
using namespace std;
void main()
{
```

```cpp
        ifstream in;
        in.open("input.txt");
        if (!in)
        {
            cout <<"Can't open the file!"<< endl;
            exit(0);
        }
        ofstream out;
        out.open("output.txt");
        if (!out)
        {
            cout<<"Can't open the file!"<<endl;
            exit(0);
        }
        char ch;
        while((ch=in.get())
            out.put(ch);
    }
```

7.2.2 字符串级读写

字串级读写，每次读或写一行(以换行符为界，因此很可能是一个段落，并非编辑器自动换行状态下的一行)，通常用于文本文件。用成员函数 getline 和插入符来读写字符串。下面的例子是逐行读入又写出汉语字符串的例子，代码如下：

```cpp
    /****************************************************************
    File name    :  f0706.cpp
    Description  :  字符串级读写
    ****************************************************************/

    #include<fstream>
    #include <iostream>
    #include<string>
    using namespace std;
    int main()
    {
        ifstream in("s.txt");
        if (!in)
        {
            cout <<"Can't open the file!"<< endl;
            exit(0);
        }
        string s;
        while (getline(in,s))
        {
            cout<<s<<endl;
        }
    }
```

假如 s.txt 的内容是：

　　C++程序设计
　　　　2016年春季

则运行的结果是：

　　c++程序设计
　　　　2016年春季

7.2.3 格式化读写

格式化读写，如读入或写出字符串表示的数值，用于文本文件。C++的格式化读写有两种方式：一是使用流类成员函数 fill、precision、setf、width 和 unsetf 等；二是使用控制符。前者需要调用这些函数，写起来麻烦一些，这里只讲控制符的用法，有的控制符不带参数，有的要带参数，见表 7-3。

表 7-3　控制符及作用

编号	控制符	作用
1	dec	输入输出时设置为十进制整数，延续性，除非重新设置
2	hex	输入输出时设置为十六进制整数，延续性，除非重新设置
3	oct	输入输出时设置为八进制整数，延续性，除非重新设置
4	setbase(n)	输入输出时设置n所指定的进制，延续性，除非重新设置
5	endl	输出换行符
6	ends	输出空白字符
7	flash	刷新输出流
8	ws	输入时跳过开头的空白字符
9	setfill(c)	输入输出时设置填充字符c(默认空格)，延续性，除非重新设置
10	setiosflags(f)	输入输出时设置f所指定的格式
11	resetiosflags(f)	输入输出时取消f所指定的格式
12	setprecision(n)	输入输出时设置n所指定的小数位数(默认6)，延续性，除非重新设置
13	setw(n)	输入输出时设置n所指定的宽度，一次性(默认0，即按本来宽度打印)

在表 7-3 中程序中 setiosflags(f) 和 resetiosflags(f) 比较特殊，现介绍如下：通过使用带参数的 setiosflags 操纵符来设置左对齐，setiosf 定义在头文件 iomanip 中。参数 ios_base::left 是 ios_base 的静态常量，因此引用时必须包括 ios_base 前缀。这里需要用 resetiosflags 操纵符关闭左对齐标志。setiosflags 不同于 width 和 setw，它的影响是持久的，直到用 resetflags 重新恢复默认值为止。具体标注和作用见表 7-4。

表 7-4　setiosflags(f) 和 resetiosflags(f) 的格式标志和作用

编号	格式标志	作用	编号	格式标志	作用
1	ios::left	输出数据左对齐	8	ios::showpoint	输出小数点和尾数
2	ios::right	输出数据右对齐(默认)	9	ios::uppercase	输出数值中E和X大写
3	ios::internal	符号位左对齐，数值右对齐	10	ios::showpos	正数显示"+"号
4	ios::dec	十进制整数	11	ios::scientific	实数用科学计数法输出
5	ios::oct	八进制整数	12	ios::fixed	实数用小数形式输出
6	ios::hex	十六进制整数	13	ios::unitbuf	刷新所有的流
7	ios::showbase	输出整数的基数	14	ios::stdio	清除标准输出流、错误流

下面的例子是从键盘读入 5 个词频,然后在屏幕上显示词频、比例和比例的负对数。输出格式都是右对齐,整数宽度 10 字节,实数用小数形式,宽度 20 字节,请注意其中控制符的用法:

```cpp
/******************************************************************
    File name    : f0707.cpp
    Description  : 格式化输出
******************************************************************/
#include <iostream>
#include <iomanip>
#include <Cmath>
using namespace std;
const int MAX_TIMES=5;
double log2(double x)                      //返回以2为底的对数
{
    return log(x)/log(2);
}
void main()
{
    const unsigned CorpusSize=100000;   //整个数据规模为100000
    int  i, freqs[MAX_TIMES];
    double ratio, logarithm;
    for(i=0; i<MAX_TIMES; i++) {
        cout << "词频: ";
        cin >> freqs[i];
    }
    cout << "\n    Frequency          ratio           -log2(ratio)\n";
    for(i=0; i<MAX_TIMES; i++)
    {
        cout << setw(10) << freqs[i];
        ratio=freqs[i]/double(CorpusSize);
        logarithm=log2(ratio);
        cout << setw(20) << setiosflags(ios::fixed) << ratio;
        cout << setw(20) << -logarithm << endl;
    }
}
```

当输入为:
词频:1000
词频:2000
词频:3000
词频:4000
词频:5000
时,输出为:

```
    Frequency          ratio           -log2(ratio)
         1000           0.010000             6.643856
```

2000	0.020000	5.643856
3000	0.030000	5.058894
4000	0.040000	4.643856
5000	0.050000	4.321928

7.2.4 二进制数据读写

二进制数据读写，可指定读写的字节个数，用于二进制文件。假定一个二进制文件中存放了若干个 WordItem 类型的词条，结构类型定义为：

```
struct WordItem
{
    char word[12];         //词汇名称
    unsigned freq;         //词汇频次
};
```

可以用流类成员函数 read 和 write 来读写一个或一批词条数据。设有：

```
WordItem a, items[Count];
ifstream in("items1.bin", ios::binary);
ofstream out("items2.bin", ios::binary);
```

从输入流当前位置读一个词条到 a：

```
in.read ((char *)&a, sizeof (WordItem));
```

将词条 a 写到输出流的当前位置：

```
out.write ((const char *)&a, sizeof(WordItem));
```

一次性读入 Count 个词条：

```
in.read ((char *) items, sizeof (WordItem)*Count);
```

一次性写出 Count 个词条：

```
out.write ((const char *) items, sizeof(WordItem)*Count);
```

根据二进制文件读取的特征，如果二进制文件里存放的全是相同类型的数据，就不必用循环控制来读入。

```
/***************************************************************
File name  : f0708.cpp
Description : 二进制输出
***************************************************************/
#include <iostream>
#include <fstream>
using namespace std;

struct  WordItem
{
        char word[12];         //词汇名称
        unsigned freq;         //词汇频次
};
```

```cpp
void main()
{
    const int Count=5;
    WordItem  items[Count],items1[Count],items2[Count];
    ofstream out1("items1.txt");
    ofstream out2("items2.bin", ios::binary);
    cout<<sizeof(WordItem)<<endl;
    string word[Count]={"is","a","the","to","in"};
    int freq[Count]={1000,2000,3000,4000,5000};
    for(int i=0;i<Count;i++)
    {
        strcpy(items[i].word,word[i].c_str());
        items[i].freq=freq[i];
        out1<<word[i].c_str()<<" "<<freq[i]<<endl;
    }
    out2.write ((const char *) items,  sizeof(WordItem)*Count);
    out1.close();
    out2.close();
    ifstream in1("items1.txt");
    ifstream in2("items2.bin", ios::binary);
    for(i=0;i<Count;i++)
    {
        in1>>items1[i].word>>items1[i].freq;
    }
    for(i=0;i<Count && !in2.eof();i++)
    {
        in2.read((char*)(items2+i),sizeof(WordItem));
    }
    for(i=0;i<Count;i++)
    {
        cout<<items1[i].word<<" "<<items1[i].freq<<endl;
        cout<<items2[i].word<<" "<<items2[i].freq<<endl;
    }
    in1.close();
    in2.close();
}
```

程序运行的结果为：

```
16
is 1000
is 1000
a 2000
a 2000
the 3000
the 3000
to 4000
to 4000
```

Item1.txt 的内容为：

```
is 1000
```

```
a 2000
the 3000
to 4000
in 5000
```

items2.bin 的内容为：

```
is 烫烫烫烫惕────────────────── a 烫烫烫烫烫？ the 烫烫烫烫？ to 烫
烫烫烫虪  in 烫烫烫烫藕
```

items2.bin 中的内容是以 2 进制方式存储的单词和词频信息。其中，单词可以看得出来，词频则由于是二进制标识，无法看出来。因此，二进制文件可读性不如文本文件。

```
for ( i=0;  i<Count  && ! in.eof ( );  i++ )
{
    in.read ( items+i, sizeof (WordItem)) ;
    ……         //检查或修改 items[i]
}
```

7.3 文件定位函数

随机存取是指通过人为地控制位置指针的具体指向，获取指定位置上的数据，或者把数据输出到文件的指定位置上。因此，调用读写函数时，文件的读写位置会自动作相应的移动。例如，读一个词条后，读写位置向输入流末尾的方向移动 sizeof(WordItem) 字节，写一个词条后，读写位置向输出流末尾的方向移动 sizeof(WordItem) 字节。判断是否到达文件末尾，是循环读写的依据，可以用流类成员函数 eof。例如，假定需要对每个词条进行检查或修改，就可以用下面的循环语句每次读入一个词条：

```
for ( i=0;  i<Count && ! in.eof ();  i++ )
{
in.read ( items+i, sizeof (WordItem));
    ……         //检查或修改 items[i]
}
```

随机读写有两个要点：一是要知道当前确切的读写位置，二是能将读写指针移动到合适的读写位置。为此提供了以下流类成员函数：tellg、tellp、seekg、seekp 和 gcount。具体函数及功能见表 7-5。

表 7-5 流类函数及主要功能

编号	函数	功能
1	gcount ()	返回最后一次输入所读入的字节数
2	tellg ()	返回输入文件指针的当前位置
3	seekg (origin)	将输入文件中指针移到指定的位置
4	seekg (offset, origin)	以参照位置为基础移动若干字节
5	tellp ()	返回输出文件指针当前的位置
6	seekp (offset, origin)	将输出文件中指针移到指定的位置
7	seekp (offset, origin)	以参照位置为基础移动若干字节

对表 7-5 说明如下：

(1) tellg 和 tellp 不带参数，容易使用。seekg 和 seekp 各带两个参数：offset 是偏移量，origin 是相对位置。

(2) offset 表示相对于 origin 移动多少字节，即要求移动到 origin+offset 处：offset 为 0，移动到 origin 处；offset 为正数，则移动到 origin 之后（朝文件末尾方向移动）；offset 为负数，则移动到 origin 之前（朝文件开头方向移动）。下面就词条输入文件举一些例子：

移到文件开头：in.seekg(0, ios::beg)；第 2 个参数可以省略
移到文件末尾：in.seekg(0, ios::end)；
移到第 i 个词条 (i 基于 0)：in.seekg(i*sizeof(WordItem), ios::beg)；
移到最后一个词条：in.seekg(-sizeof(WordItem), ios::end)；
移到上一个词条：in.seekg(-sizeof(WordItem), ios::cur)；
移到下一个词条：in.seekg(sizeof(WordItem), ios::cur)；

下面的演示程序先显示词条数组并写入文件，接着让用户修改第 3 个词条，程序直接在文件中更新这个词条的数据，然后重新显示：

```cpp
/*****************************************************************
    File name     : f0709.cpp
    Description   : 文件的定位
*****************************************************************/

#include <fstream>
#include <iostream>
#include <iomanip>
using namespace std;
const int Count=5;

struct WordItem
{
    char word[12];
    unsigned freq;
} a, items[Count] = {"boy", 10, "click", 5, "literature", 2, "friend", 21, "human", 3};

void main()
{
    fstream fp("items.bin", ios::in|ios::out|ios::binary);  //既读又写的文件
    if(!fp)
    {
        cout << "can not create file\n";
        return;
    }
    for(int i=0; i<Count; i++)
    {   //先显示一遍
        cout << setw(15) << items[i].word;
        cout << setw(10) << items[i].freq << endl;
```

```
        }
        fp.write((const char *)items, sizeof(WordItem)*Count);    //全部写入文件

        cout << "replace item 3 (word and frequency) with :";
        cin >> items[2].word >> items[2].freq;
        fp.seekp(sizeof(WordItem)*2);              //定位到第3个词条准备写,用seekp
        fp.write((const char *)(items+2), sizeof(WordItem));     //改写它

        fp.seekg(0);      //回到文件开头准备读,注意是用seekg而非seekp
        for(i=0; i<Count && !fp.eof(); i++) //再显示一遍以核对修改情况
        {
            fp.read((char *)&a, sizeof(WordItem));
            cout << setw(15) << a.word;
            cout << setw(10) << a.freq << endl;
        }
        fp.close();
    }
```

7.4 成批文件的处理

通过上述文件操作的学习,已经可以对某一个文件单独进行打开、关闭、读取、写入等各种操作。在现实的应用中,往往需要对多个名称不同的文件进行操作。例如,对 n 个文件进行词频统计,当 n 很大时,把对单个文件做词频统计的程序运行 n 次,不仅不方便而且容易出现统计错误。针对这一需求,最好的应对方式是把文件处理方法写成一个函数,然后在程序中通过循环调用这个函数,以达到处理成批文件的目的,这正是批文件处理需要解决的问题。

7.4.1 文件名骨架的设计

处理文件时必须给出文件名,但处理成批文件时,如果逐一给出文件的名称是一个非常麻烦的事情,这时可以只给出一个文件名骨架(skeleton),其中含有通配符"*"或"?",具体含义见表 7-6。

表 7-6 文件名骨架中通配符含义

编号	通配符名称	含义
1	*.txt	匹配所有扩展名为 txt 的文件
2	file.*	匹配所有名为 file 的文件,扩展名任意
3	*.*	匹配所有文件
4	file0?.txt	匹配所有扩展名为 txt 且名字前 5 个字符为 file0 的文件
5	file0?.*	匹配所有名字前 5 个字符为 file0 的文件,扩展名任意

7.4.2 库函数 _findfirst 和 _findnext

实现成批文件处理的功能,需要使用如下 2 个库函数:_findfirst 和 _findnext。它们的原型为:

```
long _findfirst(char *filespec, struct _finddata_t *fileinfo);
int _findnext(long handle, struct _finddata_t *fileinfo);
```

其中 filespec 表示文件名骨架，因为只能在当前目录中搜索，所以不能含路径。如果需要在别的目录中搜索，就应该先改变当前目录。fileinfo 是一个结构指针，所指结构类型（定义在 io.h 中）为：

```
struct _finddata_t
{
    unsigned    attrib;              //文件属性
    time_t      time_create;         //文件创建时间
    time_t      time_access;         //文件访问时间
    time_t      time_write;          //文件修改时间
    _fsize_t    size;                //文件长度
    char        name[260];           //文件名(不含路径)
};
```

在 struct _finddata_t 中，特别说明一下"文件属性"，io.h 中为此定义了以下符号常量：
_A_NORMAL：普通文件（读写不受限制）。
_A_RDONLY：只读文件（不能改写）。
_A_HIDDEN：隐藏的文件或目录。
_A_SYSTEM：系统文件或目录。
_A_SUBDIR：子目录（下层目录）。
_A_ARCH：归档文件（打算备份的）。

其中，函数 _findfirst 返回一个长整数，-1 表示搜索失败（参数 filespec 不正确，或者没有找到匹配的文件名），其他值则表示找到了与文件名骨架相匹配的第一个文件，该文件相关信息存储在 fileinfo 所指的结构变量中。这个返回值是一个编号（handle），表示当前正在进行的这一次搜索。函数 _findnext 用这个编号作为第一参数，在当前目录中继续搜索与文件名骨架相匹配的其他文件。因此，这两个函数要配合使用，步骤如下：先调用 _findfirst，对返回值进行判断，如果值不为-1，则循环调用 _findnext，处理每个匹配到的文件，直到 _findnext 的返回值不为 0。调用结束后，再调用库函数 _findclose，其目的是清除本次搜索，使下一次搜索能顺利进行。函数 _findclose 原型如下：

```
int _findclose( long handle);
```

7.4.3 批处理文件函数构建

成批文件处理的问题，关键在于两点：一是给出单个文件处理的函数，二是给出文件搜索的框架。然而，"处理"只是一个概括的说法，针对不同的具体目标，可以给出各种各样的单个文件处理的函数，如行数统计、词频统计、均值计算、自动分词等。在上述各种函数的基础上，可以编写一个专门用于成批文件处理的函数，原型如下："int process_files (const char *skeleton, void (*proc) (const char *filename))"。返回值表示处理了多少个文件。参数 skeleton 是文件名骨架，并且允许含有路径。如果含有路径，就从该参数中分解出路径（目录名字），先进入到指定的目录，否则只搜索当前目录。参数 proc 是一个无类型函数指针，指向用于处理单个文件的函数，处理单个文件的函数本身带一个文件名作为参数。于是，处理成批文件时，程序员只需编写特定的单个文件处理函数，然后用该函数的名字作为参数调用 process_files 即可。例如，假定已经定义函数"void file_count (const char *fname)"。

该函数用来统计单个文件中的字符数。现在要在目录"D:\MyDir"中对所有文本文件进行行数统计，就可以这样来调用 process_files：

```
int file_num = process_files ("C:\\Mydir\\*.txt", line_count);
```

具体 process_files 的处理文件函数定义如下：

```cpp
/*******************************************************************
File name    : f0710.cpp
Description  : 成批文件的处理
*******************************************************************/
#include <fstream>
#include <iostream>
#include <strstream>
#include <direct.h>
#include <io.h>
using namespace std;
int process_files(const char *skeleton, void (*proc) (const char *filename))
{
    unsigned files=0;
    long handle;
    struct _finddata_t info;
    char buffer[100], *p;
    strcpy(buffer, skeleton);
    p=strrchr(buffer, '\\');
    if(p!=NULL)                              //如果含有路径
    {
        char ch = *(++p);
        *p = '\0';
        if(_chdir(buffer)!=0) return 0;
        *p = ch;
    }
    else p=buffer;
    handle=_findfirst(p, &info);
    if(handle == -1) return 0;
    do {
        proc(info.name);                     //调用 proc 所指的函数
        files++;
    } while(_findnext(handle, &info)==0);
    return files;
}
void file_count(const char *filename)
{
    char ch;
    int count=0;
    ifstream in(filename);
    if (!in)
    {
        cout<<"Can't open the file!"<<endl;
        exit(0);
```

```
        }
        while((ch=in.get())!=EOF)
        {
            cout<<int(ch);
            if(ch!=' '&&ch!='\r'&&ch!='\n')
            {
                cout<<'('<<ch<<')';
                count++;
            }
            cout<<' ';
        }
        printf("\nthe file <%s> has %d characters\n",filename,count);
}
void main()
{
    process_files("c:\\mydir\\*.txt",file_count);
}
```

假如 c 盘的 mydir 文件夹有三个文件，内容分别为：

```
1.txt
1 2345
2.txt
12345  (注意，5 后面有一个空格)
s.txt
1 2345
12345  (注意，5 后面有一个空格)
```

则运行的结果是：

```
49(1) 32 50(2) 51(3) 52(4) 53(5)
the file <1.txt> has 5 characters
49(1) 50(2) 51(3) 52(4) 53(5) 32
the file <2.txt> has 5 characters
49(1) 32 50(2) 51(3) 52(4) 53(5) 10 49(1) 50(2) 51(3) 52(4) 53(5) 32
the file <s.txt> has 10 characters
```

在定义的批处理函数中 proc 所指的函数是定义带函数指针的函数的关键。proc 是一个形参，真正调用的是实参，即该指针所指的函数，如字符统计的 file_count，词频统计的 word_count 或文本切词的 segmentation 等，process_files 只是把文件名(info.name)传给被调用的函数而已。从实参到形参是一种概括过程，这种概括带来的好处就是带函数指针的函数的代码重用。在 process_files 定义中被调用的函数可能有一些输出数据，如函数 file_count 统计出单个文件中的字符数，应该如何处理呢？有多种处理办法：可以直接送到标准输出，也可以在调用 process_files 的程序中设置一个全局变量以统计全部文件的行数。当然，还有一种办法，就是让 file_count 函数有返回值，或者有一个输出参数。

注意：这里判断文件结束没有用 eof()。加入改成用 eof()判断，就是将 while(in.peek()!=EOF)改成 while(!in.eof())，运行的结果为：

```
1 23455the file <1.txt> has 7 characters
12345the file <2.txt> has 7 characters
```

```
1 2345
12345 the file <s.txt> has 14 characters
```

7.5 文件操作程序举例

设有一个文本文件,其名称为"待统计分词文件",该文件中均为已经分过词的汉语词汇序列,如果该文件具体存储如下文本内容"2015年8月,中国推出中国人民抗日战争暨世界反法西斯战争胜利70周年纪念币和中国人民抗日战争暨世界反法西斯战争胜利70周年纪念邮票。"请统计每一个文本中每一个汉语词汇的频次。

参考源程序如下:

```cpp
/*******************************************************************
    File name   : f0711.cpp
    Description : 单文件词语频次统计处理
*******************************************************************/
#include <iostream>
#include <strstream>
#include <fstream>
#include <io.h>
#include <string>
using namespace std;
void main()
{
    ifstream in("待统计分词文件.txt");
    if (!in)
    {
        cout <<"Can't open the file!"<< endl;
        exit(0);
    }
    ofstream out("词频统计结果.txt");
    string line,w;
    int count=0;
    while(getline(in,line))                    //循环读取每一行
    {
        if(line=="") continue;
        line+=" ";
        istrstream sline(line.c_str());        //将每一行内容传递给流变量
        while(true)
        {
            sline >> w;          //读取流变量中的每一个词语(空格或制表符间隔)
            count++;             //每读取一次,词频加一
            if(sline.eof()) break;
        }
    }
    out << "共有" << count << "个词语!" << endl;
    cout << "统计完毕!" << endl;
}
```

设有两个文本文件,其名称为"待统计分词文件1"和"待统计分词文件2",两个文件中均为已经分过次的汉语词汇序列,如果文件1具体存储如下文本内容:"2015年8月,中国

推出中国人民抗日战争暨世界反法西斯战争胜利70周年纪念币和中国人民抗日战争暨世界反法西斯战争胜利70周年纪念邮票。",文件2存储如下内容:"习近平总书记在纪念中国人民抗日战争暨世界反法西斯战争胜利70周年的系列重要讲话引发社会各界高度关注。"请设计批处理函数完成对两个文本中每一个汉语词汇的频次。

参考源程序如下:

```cpp
/******************************************************************
    File name    : f0712.cpp
    Description : 多文件词语频次统计处理
******************************************************************/

#include <iostream>
#include <strstream>
#include <fstream>
#include <io.h>
#include <string>
using namespace std;

void main()
{
    long  handle;                               //批处理句柄
    struct _finddata_t  info;                   //批处理文件的信息存放到info结构体中
    handle=_findfirst("*.txt", &info);          //批处理文件的后缀为"txt",文件信息
                                                //存放到info结构体中

    if(handle == -1)
        return;
    do
    {
        char output_name[100];
        strcpy(output_name,info.name);
        strcat(output_name,".词频统计结果");
        ifstream in(info.name);
        if (!in)
        {
            cout <<"Can't open the file!"<< endl;
            exit(0);
        }
        ofstream out(output_name);
        if (!out)
        {
            cout<<"Can't open the file!"<<endl;
            exit(0);
        }
        string line,w;
        int count=0;
        while(getline(in,line))
```

```
            {
                if(line=="")
                    continue;
                line+=" ";
                istrstream sline(line.c_str());
                while(true)
                {
                    sline >> w;
                    count++;
                    if(sline.eof())
                        break;
                }
            }
            out << "共有" << count << "个词语!" << endl;
        } while(_findnext(handle, &info)==0);
                                //当目录下没有符合后缀要求的文件时,循环结束
        cout << "统计完毕!" << endl;
        return;
}
```

练 习

现有三个文本文件,文件名称分别为 text1、text2 和 text3,文件中分别存放了如下的内容:

"中新网 10 月 5 日电据诺贝尔奖官网的最新消息,瑞典斯德哥尔摩当地时间 5 日中午 11 时 30 分,2015 年诺贝尔生理学或医学奖在当地的卡罗琳斯卡医学院揭晓,爱尔兰医学研究者威廉·坎贝尔、日本学者 Satoshi Omura 以及中国药学家屠呦呦荣获了该奖项。"

"在莫言的家乡,富饶的德行一直都在与最邪恶的残忍交战,对那些有勇气闯进去一窥其究竟的人士来说,所面临的将是一次步履艰难的文学冒险之行。中国,乃至世界的其他地方,何曾经受过这样一种史诗春潮的波澜冲击?在莫言的作品中,"世界文学"发出了让众多的当代人倾倒折服的声音。"

"鲁迅先生说,要改造国人的精神世界,首推文艺。举精神之旗、立精神支柱、建精神家园,都离不开文艺。当高楼大厦在我国大地上遍地林立时,中华民族精神的大厦也应该巍然耸立。"

1. 设计一个 C++程序,读取 text1、text2 和 text3 三个文本文件中的内容,把上述三个文本文件中的内容输出到屏幕上。

2. 选择一种字符编码,设计一个 C++程序,该程序的主要功能为统计三个文件中任一文件中的汉字出现的频次(英文单词在此处被作为一个汉字进行统计)并把统计结果输入到文件名为 "单个文件汉字字频统计结果.txt" 的文件中。

3. 设计一个 C++程序,其中定义一个子函数完成对单个文本文件中汉字字频的统计,在主函数中调用成批文件处理函数以及具有汉字字频统计功能的子函数,完成对 text1、text2 和 text3 三个文本文件的汉字字频统计,并把统计结果输出到 "多个文件汉字字频统计结果.txt" 的文件中。

第8章 面向对象程序设计应用举例

面向对象程序设计方法将软件开发抽象成一些不同类型的对象集合，然后像搭积木一样进行无依赖的设计，从而达到最大化的软件重用和软件开发效率，这就是抽象编程。使用抽象类做界面，可以彻底分离编程内容和职责，建立完全独立的模块，而且在界面不变的情况下，独立模块的变化丝毫不影响应用程序的运行。

本章将以一个简单的图形管理程序为例，介绍面向对象程序设计方法及其应用。

- **知识要点**

面向对象基本分析方法；

面向对象程序设计方法：类、继承、多态等知识。

- **探索思考**

当面对一个问题时，如何分析问题？如何确定问题涉及的类，以及类之间的关系？

- **预备知识**

类的知识、继承、多态、抽象编程

8.1 问题提出

在很多应用系统里，需要管理很多不同的图形。例如：在一个钢板配料系统中，根据不同用途要求在一个大小确定的钢板材料上，最大化利用钢板材料，将钢板材料根据配料结果裁剪成不同形状的块，如何描述和管理这些钢板材料块，是一个配料系统中必须考虑的问题。再如，固定空间物品装载系统，也需要根据物品形状，最大化地利用空间，一旦确定了装载方案，则每个形状物品的位置就需要精确描述。如何描述和管理这些物品位置也是最优装载系统必须考虑的问题。此类问题，在实际应用系统中经常遇到。下面将用面向对象的方法来分析此类问题的解决方法。

首先假设管理的图形主要有线段、圆、三角形、矩形、立方体、圆柱体等，要求设计一个程序有效管理这些图形。这些图形主要有以下特性。

- 线段

属性特征：有一个起点和一个终点。

行为特征：获取线段长度、平移、显示、获取起点和终点坐标。

- 圆

属性特征：圆心和半径。

行为特征：计算圆面积、周长、平移、显示、获取圆心坐标。

- 三角形

属性特征：顶点。

行为特征：面积、平移、显示、获取顶点坐标、显示。

- 矩形

属性特征：两个二维顶点。

行为特征：面积、平移、周长、获取顶点坐标、显示。

- 立方体

属性特征：两个三维顶点。

行为特征：体积、表面积、平移、获取订单坐标、显示。

- 圆柱体

属性特征：圆心、半径和高。

行为特征：体积、表面积、平移、获取圆心坐标、半径和高、显示。

8.2 系统设计

从 8.1 节提出的各类图形特性，可以抽象出圆类、三角形类、矩形类、立方体类、圆柱体类，还有一个线段类。仔细思考可以发现，线段一般包含一维数轴上的线段、二维平面中的线段和三维空间的线段，并且三类线段计算长度的方法也不同，因此，可以将线段再抽象为三类，分别是：一维数轴上的线段、二维平面中的线段和三维空间的线段。其次，每个线段都由两个端点确定，而且圆、三角形等也都有端点数据，可以再抽象出一个点类，以作为其他类的数据成员。通用点类也可以分为一维数轴上的点、二维平面中的点和三维空间的点。

从一维点和一维线段分析，这两类都有如下共同的行为特征：显示对象属性、平移。当一维点看成是一条长度为 0 的线段时，一维点和一维线段还有一个共性的行为特征就是获取长度。因此，可以对一维点和一维线段再进行抽象，得到一维图形抽象类。

对二维点、二维线段、圆、三角形、矩形进行分析，这些类型图形具有共同的行为特征：显示、平移、计算面积。二维点和二维线段的面积设为 0 的情况下，也可以计算面积，这样便于统一不同类型之间的共同行为特征。因此，可以再次对这些类型对象进行抽象，得到二维图形抽象类，拥有的行为特征：显示、平移、计算面积。

同样对三维点、立方体和圆柱体进行分析，这些类图形具有共同行为特征：显示、平移、计算面积和计算体积。此处也假设三维点的面积和体积为 0。因此，再次对这些类型对象进行抽象，得到三维图形抽象类，拥有的行为特征：显示、平移、计算面积和计算体积。

一维图形抽象类、二维图形抽象类和三维图像抽象类都有显示行为特征，这里对三类抽象类再次抽象得到一个图形抽象基类，拥有的行为特征：显示。

根据以上分析，最终可以得到类的层次结构如图 8-1 所示。

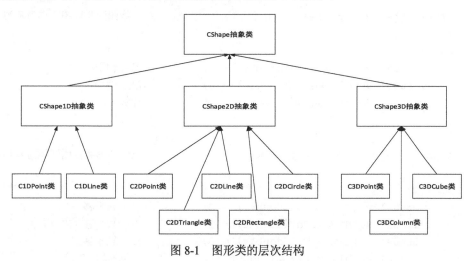

图 8-1 图形类的层次结构

使用抽象类做界面既可以彻底分离程序模块，又可以克服烦琐的成员函数转换，干净又利落。在彻底分离了编程内容和职责之后，就可以建立完全独立的模块。只有独立程序模块的外部接口界面不发生改变，则程序模块的内部实现发生变化丝毫不会影响到用户应用程序的运行，从而做到了用户与类代码的完全分离。只要界面确定了，二者可以毫无干涉的独立开发软件。

8.3 系统实现

图 8-1 所示的图形类层次结构的类定义代码如下：

```
/***************************************************************
    File name    : shape.h
    Description : 图形抽象基类头文件
***************************************************************/
#ifndef _SHAPE_H
#define _SHAPE_H

const double pi = 3.14158;          //定义全局常量 pi
class CShape                        //图形类是一个抽象基类，为派生类提供统一界面
{
public:
    virtual void show()const = 0;   //纯虚函数
};

class CShape1D : public CShape              //一维图形类以图形类为基类
{
public:
    virtual void MoveTo(double) = 0;        //平移
    virtual double GetLength()const = 0;    //计算长度
};


class CShape2D :public CShape               //二维图形类以图形类为基类
{
public:
    virtual void MoveToX(double) = 0;       //沿 X 轴平移
    virtual void MoveToY(double) = 0;       //沿 Y 轴平移
    virtual double GetArea()const = 0;      //计算面积
};

class CShape3D :public CShape               //三维图形类以图形类为基类
{
public:
    virtual void MoveToX(double) = 0;       //沿 X 轴平移
    virtual void MoveToY(double) = 0;       //沿 y 轴平移
    virtual void MoveToZ(double) = 0;       //沿 Z 轴平移
    virtual double GetArea()const = 0;      //计算面积
```

```cpp
    virtual double GetVolume()const = 0;         //计算体积
};
#endif
//================================================================
/*********************************************************************
    File name   : 1DPoint.h
    Description : 1 维点类头文件
*********************************************************************/
#ifndef _1D_POINT_H
#define _1D_POINT_H

#include "shape.h"

class C1DPoint : public CShape1D
{
private:
    double m_x;                                  //点坐标
public:
    C1DPoint(double = 0);                        //构造函数
    virtual void MoveTo(double x);               //平移
    virtual double GetLength()const;             //计算长度
    virtual void show()const;                    //输出函数
    double GetX()const;                          //获取 x 坐标
    double GetDistance(const C1DPoint&)const;    //计算两个点之间的距离

    /*friend function------------*/
    friend int operator==(const C1DPoint&, const C1DPoint&);
                                                 //比较两个点是否相同
};
#endif
//================================================================
/*********************************************************************
    File name   : 1DPoint.cpp
    Description : 1 维点类源文件
*********************************************************************/
#include <iostream>
#include "1DPoint.h"
using namespace std;
/*member function ----------------------------------------------*/
C1DPoint::C1DPoint(double x)
{
    m_x = x;
}
void C1DPoint::MoveTo(double x)
{
    m_x += x;
}
double C1DPoint::GetLength()const
```

```cpp
{
    return 0;
}
double C1DPoint::GetX()const
{
    return m_x;
}
void C1DPoint::show()const
{
    cout << m_x;
}
double C1DPoint::GetDistance(const C1DPoint& point)const   //计算两个点之间的距离
{
    return m_x - point.m_x;
}
/*friend function-------------*/
int operator==(const C1DPoint& point1, const C1DPoint& point2)
{
    if (point1.m_x == point2.m_x)
        return 1;
    else
        return 0;
}
//================================================================

/***********************************************************************
    File name    : 1DLine.h
    Description : 1 维线类头文件
***********************************************************************/
#ifndef _1D_LINE_H
#define _1D_LINE_H
#include "shape.h"
#include "1DPoint.h"
class C1DLine :public CShape1D
{
private:
    C1DPoint m_point1, m_point2;
public:
    C1DLine(double = 0, double = 1);                            //构造函数
    C1DLine(const C1DPoint&, const C1DPoint&);                  //构造函数
    virtual void MoveTo(double x);                              //平移
    virtual double GetLength()const;                            //计算长度
    virtual void show()const;                                   //输出函数
    C1DPoint GetPoint(int index)const;                          //获取第一个坐标点
};
#endif
//================================================================
/***********************************************************************
```

```cpp
    File name    : 1DLine.cpp
    Description  : 1 维线类源文件
****************************************************************/
#include <iostream>
#include "1DLine.h"
using namespace std;
/*member function ---------------------------------------------*/
C1DLine::C1DLine(double x1, double x2) :m_point1(x1), m_point2(x2)
{
    if (x1 == x2)
        throw string("1DLine"); //如果构成线的两个点相同,则不构成线,抛出异常
}
C1DLine::C1DLine(const C1DPoint& point1, const C1DPoint& point2) :
                m_point1(point1), m_point2(point2)
{
    if (point1 == point2)
        throw string("1DLine"); //如果构成线的两个点相同,则不构成线,抛出异常
}
void C1DLine::MoveTo(double x)
{
    m_point1.MoveTo(x);
    m_point2.MoveTo(x);
}
double C1DLine::GetLength()const
{
    return m_point2.GetDistance(m_point1);
}
C1DPoint C1DLine::GetPoint(int index)const
{
    return index == 0 ? m_point1 : m_point2;
}
void C1DLine::show()const
{
    m_point1.show();
    cout << "---->";
    m_point2.show();
}
//================================================================
/****************************************************************
    File name    : 2DPoint.h
    Description  : 2 维点类头文件
****************************************************************/
#ifndef _2D_POINT_H
#define _2D_POINT_H
#include "shape.h"
class C2DPoint :public CShape2D
{
private:
```

```cpp
    double m_x, m_y;                                    //坐标
public:
    C2DPoint(double x = 0, double y = 0);
    double GetX()const;
    double GetY()const;
    double GetDistance(const C2DPoint&)const;
    virtual void MoveToX(double);                       //沿 X 轴平移
    virtual void MoveToY(double);                       //沿 Y 轴平移
    virtual double GetArea()const;                      //计算面积
    virtual void show()const;

    /*friend function --------------*/
    friend int operator==(const C2DPoint&, const C2DPoint&); //比较两个点是否相同
};
#endif
//======================================================================

/**********************************************************************
    File name    : 2DPoint.cpp
    Description  : 2 维点类源文件
**********************************************************************/
#include <iostream>
#include "2DPoint.h"
using namespace std;
/*member function---------------------------------------------------*/
C2DPoint::C2DPoint(double x, double y)
{
    m_x = x;
    m_y = y;
}
double C2DPoint::GetX()const
{
    return m_x;
}
double C2DPoint::GetY()const
{
    return m_y;
}
double C2DPoint::GetDistance(const C2DPoint& point)const
{
    return sqrt(((m_x - point.m_x)*(m_x - point.m_x)) + ((m_y -
            point.m_y)*(m_y - point.m_y)));
}
void C2DPoint::MoveToX(double x)
{
    m_x += x;
}
void C2DPoint::MoveToY(double y)
```

```cpp
{
    m_y += y;
}
double C2DPoint::GetArea()const
{
    return 0;
}
void C2DPoint::show()const
{
    cout << "(" << m_x << "," << m_y << ")";
}
/*friend function --------------*/
int operator==(const C2DPoint& point1, const C2DPoint& point2)
{
    if ((point1.m_x == point2.m_x) && (point1.m_y == point2.m_y))
        return 1;
    else
        return 0;
}
//=====================================================================
/*********************************************************************
    File name    : 2DLine.h
    Description  : 2维线类头文件
*********************************************************************/
#ifndef _2D_LINE_H
#define _2D_LINE_H
#include "shape.h"
#include "2DPoint.h"
class C2DLine :public CShape2D
{
private:
    C2DPoint m_point1, m_point2;
public:
    C2DLine(double = 0, double = 0, double = 0, double = 0);
    C2DLine(const C2DPoint&, const C2DPoint&);
    C2DPoint GetPoint(int);
    double GetLength()const;
    virtual void MoveToX(double);         //沿 X 轴平移
    virtual void MoveToY(double);         //沿 Y 轴平移
    virtual double GetArea()const;        //计算面积
    virtual void show()const;             //显示线信息，虚函数
};
#endif
//=====================================================================
/*********************************************************************
    File name    : 2DLine.cpp
    Description  : 2维线类源文件
*********************************************************************/
```

```cpp
#include <iostream>
#include "2DLine.h"
using namespace std;
/*member function ---------------------------------------------------*/
C2DLine::C2DLine(double x1, double y1, double x2, double y2) :m_point1(x1, y1),
            m_point2(x2, y2)
{
    if ((x1 == x2) && (y1 == y2))
        throw string("2DLine");       //构建线段失败，抛出异常
}
C2DLine::C2DLine(const C2DPoint& point1, const C2DPoint& point2)
:m_point1(point1), m_point2(point2)
{
    if (point1 == point2)
        throw string("2DLine");       //构建线段失败，抛出异常
}
C2DPoint C2DLine::GetPoint(int index)
{
    return index == 0 ? m_point1 : m_point2;
}
double C2DLine::GetLength()const
{
    return m_point1.GetDistance(m_point2);
}
void C2DLine::MoveToX(double x)      //沿 X 轴平移
{
    m_point1.MoveToX(x);
    m_point2.MoveToX(x);
}
void C2DLine::MoveToY(double y)      //沿 Y 轴平移
{
    m_point1.MoveToY(y);
    m_point2.MoveToY(y);
}
double C2DLine::GetArea()const       //计算面积
{
    return 0;
}
void C2DLine::show()const
{
    m_point1.show();
    cout << "--->";
    m_point2.show();
}
//================================================================
/*****************************************************************
    File name   : 2DCircle.h
    Description : 2 维圆类头文件
```

```cpp
****************************************************************/
#ifndef _2D_CIRCLE_H
#define _2D_CIRCLE_H
#include "shape.h"
#include "2DPoint.h"
class C2DCircle:public CShape2D
{
private:
    C2DPoint m_point;                   //圆心坐标
    double m_r;                         //半径
public:
    C2DCircle(double = 0, double = 0, double = 1);
    C2DCircle(const C2DPoint&, double);
    C2DPoint GetPoint()const;           //获取圆心坐标
    double GetRadius()const;            //获取半径
    virtual void MoveToX(double);       //沿X轴平移
    virtual void MoveToY(double);       //沿Y轴平移
    virtual double GetArea()const;      //计算面积
    virtual void show()const;           //显示圆信息，虚函数
};
#endif
//========================================================================
/*********************************************************************
    File name    : 2DCircle.cpp
    Description : 2维圆类源文件
****************************************************************/
#include <iostream>
#include "2DCircle.h"
using namespace std;
/*member function --------------------------------------------------*/
C2DCircle::C2DCircle(double x, double y, double r):m_point(x,y)
{
    if (0 == r)
        throw string("Circle");         //如果半径为0，则不构成圆，抛出异常
    m_r = r;
}
C2DCircle::C2DCircle(const C2DPoint& point, double r) : m_point(point)
{
    if (0 == r)
        throw string("error");          //如果半径为0，则不构成圆，抛出异常
    m_r = r;
}
C2DPoint C2DCircle::GetPoint()const     //获取圆心坐标
{
    return m_point;
}
double C2DCircle::GetRadius()const      //获取半径
{
```

```cpp
        return m_r;
}
void C2DCircle::MoveToX(double x)           //沿 X 轴平移
{
    m_point.MoveToX(x);
}
void C2DCircle::MoveToY(double y)           //沿 Y 轴平移
{
    m_point.MoveToY(y);
}
double C2DCircle::GetArea()const            //计算面积
{
    return pi*m_r*m_r;
}
void C2DCircle::show()const
{
    m_point.show();
    cout << "-----" << m_r;
}
//================================================================
/*****************************************************************
    File name   : 2DRectangle.h
    Description : 2 维矩形类头文件
*****************************************************************/
#ifndef _3D_RECTANGLE_H
#define _3D_RECTANGLE_H
#include "2DPoint.h"
class C2DRectangle :public CShape2D
{
private:
    C2DPoint m_point1, m_point2;
public:
    C2DRectangle(double = 0, double = 0, double = 1, double = 1);
    C2DRectangle(const C2DPoint&, const C2DPoint&);
    C2DPoint GetPoint(int)const;                //获取点
    virtual void MoveToX(double);               //沿 X 轴平移
    virtual void MoveToY(double);               //沿 Y 轴平移
    virtual double GetArea()const;              //计算面积
    virtual void show()const;                   //显示信息，虚函数
};
#endif
//================================================================
/*****************************************************************
    File name   : 2DRectangle.cpp
    Description : 2 维矩形类源文件
*****************************************************************/
#include <iostream>
#include "2DRectangle.h"
```

```cpp
using namespace std;
/*member function -----------------------------------------------------*/
C2DRectangle::C2DRectangle(double x1, double y1, double x2, double y2)
        :m_point1(x1, y1), m_point2(x2, y2)
{
    if ((x1 == x2) || (y1 == y2))
    {
        throw string("Rectangle");
    }
}
C2DRectangle::C2DRectangle(const C2DPoint& point1,
const C2DPoint& point2) :m_point1(point1), m_point2(point1)
{
    if (m_point1 == m_point2)
    {
        throw string("Rectangle");
    }
}
C2DPoint C2DRectangle::GetPoint(int index)const          //获取点
{
    return index == 0 ? m_point1 : m_point2;
}
void C2DRectangle::MoveToX(double x)                     //沿 X 轴平移
{
    m_point1.MoveToX(x);
    m_point2.MoveToX(x);
}
void C2DRectangle::MoveToY(double y)                     //沿 Y 轴平移
{
    m_point1.MoveToY(y);
    m_point2.MoveToY(y);
}
double C2DRectangle::GetArea()const                      //计算面积
{
    double d1 = m_point1.GetX() - m_point2.GetX();
    double d2 = m_point1.GetY() - m_point2.GetY();
    return d1*d2;
}
void C2DRectangle::show()const
{
    m_point1.show();
    cout << "=====>";
    m_point2.show();
}
//====================================================================
/********************************************************************
    File name    : 2DTriangle.h
    Description  : 2 维三角形类头文件
```

```cpp
**********************************************************************/
#ifndef _2D_TRIANGLE_H
#define _2D_TRIANGLE_H
#include "2DPoint.h"
class C2DTriangle :public CShape2D
{
private:
    C2DPoint m_point1, m_point2, m_point3;
public:
    C2DTriangle(double = 0, double = 0, double = 1, double = 1, double = 2,
            double = 2);
    C2DTriangle(const C2DPoint& point1, const C2DPoint& point2, const
            C2DPoint& point3);
    C2DPoint GetPoint(int);              //获取点
    virtual void MoveToX(double);        //沿X轴平移
    virtual void MoveToY(double);        //沿Y轴平移
    virtual double GetArea()const;       //计算面积
    virtual void show()const;
};
#endif
//====================================================================
/*********************************************************************
    File name   : 2DTriangle.h
    Description : 2维三角形类头文件
**********************************************************************/
#include <iostream>
#include <string>
#include "2DTriangle.h"
using namespace std;

/*member function ---------------------------------------*/
C2DTriangle::C2DTriangle(double x1, double y1, double x2, double y2,
        double x3, double y3): m_point1(x1, y1), m_point2(x2, y2),
        m_point3(x3,y3)
{
    if (((x1 - x2) / (y1 - y2)) == ((x2 - x3) / (y2 - y3)))
        throw string("Triangle");
}
C2DTriangle::C2DTriangle(const C2DPoint& point1, const C2DPoint& point2,
                const C2DPoint& point3): m_point1(point1),
                m_point2(point2), m_point3(point3)
{
    double x1 = m_point1.GetX() - m_point2.GetX();
    double y1 = m_point1.GetY() - m_point2.GetY();
    double x2 = m_point2.GetX() - m_point3.GetX();
    double y2 = m_point2.GetY() - m_point3.GetY();
    if ( x1 /y1  == x2/y2 )
```

```cpp
        {
            throw string("Triangle");
        }
}
C2DPoint C2DTriangle::GetPoint(int index)        //获取点
{
    if (index == 0)
        return m_point1;
    else if (index == 1)
        return m_point2;
    else
        return m_point3;
}
void C2DTriangle::MoveToX(double x)              //沿 X 轴平移
{
    m_point1.MoveToX(x);
    m_point2.MoveToX(x);
    m_point3.MoveToX(x);
}
void C2DTriangle::MoveToY(double y)              //沿 Y 轴平移
{
    m_point1.MoveToY(y);
    m_point2.MoveToY(y);
    m_point3.MoveToY(y);
}
double C2DTriangle::GetArea()const               //计算面积
{
    double s1, s2, s3;                           //三条边长
    s1 = m_point1.GetDistance(m_point2);
    s2 = m_point2.GetDistance(m_point3);
    s3 = m_point3.GetDistance(m_point1);
    double s = (s1 + s2 + s3) / 2;
    double area=sqrt(s*(s-s1)*(s-s2)*(s-s3));
    return area;
}
void C2DTriangle::show()const
{
    m_point1.show();
    cout << "---->";
    m_point2.show();
    cout << "---->";
    m_point3.show();
    cout << "---->";
    m_point1.show();
}
//===================================================================
/*******************************************************************
    File name   : 3DPoint.h
```

```cpp
        Description : 3 维点类头文件
***************************************************************/
#ifndef _3D_POINT_H
#define _3D_POINT_H
#include "shape.h"
class C3DPoint :public CShape3D
{
private:
    double m_x, m_y, m_z;
public:
    C3DPoint(double = 0, double = 0, double = 0);
    double GetX()const;
    double GetY()const;
    double GetZ()const;
    double GetDistance(const C3DPoint&)const;
    virtual void MoveToX(double);         //沿 X 轴平移
    virtual void MoveToY(double);         //沿 Y 轴平移
    virtual void MoveToZ(double);         //沿 Z 轴平移
    virtual double GetArea()const;        //计算面积
    virtual double GetVolume()const;      //计算体积
    virtual void show()const;

    /*friend function --------------------*/
    friend int operator==(const C3DPoint&, const C3DPoint&);
};
#endif
//==================================================================

/******************************************************************
    File name   : 3DPoint.h
    Description : 3 维点类头文件
***************************************************************/
#include <iostream>
#include "3DPoint.h"
using namespace std;
/*member function -------------------------------------------------*/
C3DPoint::C3DPoint(double x, double y, double z)
{
    m_x = x;
    m_y = y;
    m_z = z;
}
double C3DPoint::GetX()const
{
    return m_x;
}
double C3DPoint::GetY()const
{
```

```cpp
        return m_y;
}
double C3DPoint::GetZ()const
{
        return m_z;
}
double C3DPoint::GetDistance(const C3DPoint& point)const
{
        return sqrt((m_x - point.m_x)*(m_y - point.m_y)*(m_z - point.m_z));
}
void C3DPoint::MoveToX(double x)            //沿 X 轴平移
{
        m_x += x;
}
void C3DPoint::MoveToY(double y)            //沿 Y 轴平移
{
        m_y += y;
}
void C3DPoint::MoveToZ(double z)
{
        m_z += z;
}
double C3DPoint::GetArea()const             //计算面积
{
        return 0;
}
double C3DPoint::GetVolume()const           //计算体积
{
        return 0;
}
void C3DPoint::show()const
{
        cout << "(" << m_x << "," << m_y << "," << m_z << ")";
}

/*friend function -------------------*/
int operator==(const C3DPoint& point1, const C3DPoint& point2)
{
        if ((point1.m_x == point2.m_x)
            && (point1.m_y == point2.m_y)
            && (point1.m_z == point2.m_z))
            return 1;
        else
            return 0;
}
//==================================================================
/*******************************************************************
        File name   : 3DCube.h
```

```
        Description : 3维立方体类头文件
*********************************************************************/
#ifndef _3D_CUBE_H
#define _3D_CUBE_H
#include "shape.h"
#include "3DPoint.h"
class C3DCube :public CShape3D
{
private:
    C3DPoint m_point1, m_point2;
public:
    C3DCube(double = 0, double = 0, double = 0, double = 1, double = 1, double = 1);
    C3DCube(const C3DPoint&, const C3DPoint&);
    C3DPoint GetPoint(int) const;
    virtual void MoveToX(double);        //沿X轴平移
    virtual void MoveToY(double);        //沿Y轴平移
    virtual void MoveToZ(double);        //沿Z轴平移
    virtual double GetArea()const;       //计算面积
    virtual double GetVolume()const;     //计算体积
    virtual void show()const;
};
#endif
//====================================================================
/*********************************************************************
    File name   : 3DCube.cpp
    Description : 3维立方体类源文件
*********************************************************************/
#include <iostream>
#include "3DCube.h"
using namespace std;

/*member function ---------------------------------------------*/
C3DCube::C3DCube(double x1, double y1, double z1, double x2, double y2, double z2)
:m_point1(x1, y1, z1), m_point2(x2, y2, z2)
{
    if ((x1 == x2) || (y1 == y2) || (z1 == z2))
        throw string("Cube");
}
C3DCube::C3DCube(const C3DPoint& point1, const C3DPoint& point2)
: m_point1(point1), m_point2(point2)
{
    if (m_point1 == m_point2)
        throw string("Cube");
}
C3DPoint C3DCube::GetPoint(int index)const
{
    return index == 0?m_point1: m_point2;
}
```

```cpp
void C3DCube::MoveToX(double x)              //沿 X 轴平移
{
    m_point1.MoveToX(x);
    m_point2.MoveToX(x);
}
void C3DCube::MoveToY(double y)              //沿 Y 轴平移
{
    m_point1.MoveToY(y);
    m_point2.MoveToY(y);
}
void C3DCube::MoveToZ(double z)              //沿 Z 轴平移
{
    m_point1.MoveToZ(z);
    m_point2.MoveToZ(z);
}
double C3DCube::GetArea()const               //计算面积
{
    /*计算表面积 ----*/
    double d1 = m_point1.GetX() - m_point2.GetX();
    double d2 = m_point1.GetY() - m_point2.GetY();
    double d3 = m_point1.GetZ() - m_point2.GetZ();

    double area = d1*d2 + d2*d3 + d3*d1;
    return 2 * area;
}
double C3DCube::GetVolume()const             //计算体积
{
    double d1 = m_point1.GetX() - m_point2.GetX();
    double d2 = m_point1.GetY() - m_point2.GetY();
    double d3 = m_point1.GetZ() - m_point2.GetZ();
    return d1*d2*d3;
}
void C3DCube::show()const
{
    m_point1.show();
    cout << "=======>";
    m_point2.show();
}
//================================================================
/****************************************************************
    File name   : 3DColumn.h
    Description : 3 维圆柱体类头文件
****************************************************************/
#ifndef _3D_COLUMN_H
#define _3D_COLUMN_H
#include "shape.h"
#include "3DPoint.h"
class C3DColumn:public CShape3D
```

```cpp
{
private:
    C3DPoint m_point;
    double m_r, m_high;
public:
    C3DColumn(double = 0, double = 0, double = 0, double = 1, double = 1);
    C3DColumn(const C3DPoint&, double, double);
    C3DPoint GetPoint() const;
    double GetR()const;
    double GetHigh()const;
    virtual void MoveToX(double);          //沿 X 轴平移
    virtual void MoveToY(double);          //沿 Y 轴平移
    virtual void MoveToZ(double);          //沿 Z 轴平移
    virtual double GetArea()const;         //计算面积
    virtual double GetVolume()const;       //计算体积
    virtual void show()const;
};
#endif
//================================================================
/******************************************************************
    File name   : 3DColumn.h
    Description : 3 维圆柱体类源文件
******************************************************************/
#include <iostream>
#include "3DColumn.h"
using namespace std;

/*member function ---------------------------------------------*/
C3DColumn::C3DColumn(double x, double y, double z, double r, double h):
                m_point(x, y, z)
{
    m_r = r;
    m_high = h;
    if ((h == 0) || (r == 0))
        throw string("Column");
}
C3DColumn::C3DColumn(const C3DPoint& point, double r, double h) :m_point(point)
{
    m_r = r;
    m_high = h;
    if ((h == 0) || (r == 0))
        throw string("Column");
}
inline C3DPoint C3DColumn::GetPoint() const
{
    return m_point;
}
inline double C3DColumn::GetR()const
```

```
    {
        return m_r;
    }
    double C3DColumn::GetHigh()const
    {
        return m_high;
    }
    void C3DColumn::MoveToX(double x)        //沿 X 轴平移
    {
        m_point.MoveToX(x);
    }
    void C3DColumn::MoveToY(double y)        //沿 Y 轴平移
    {
        m_point.MoveToY(y);
    }
    void C3DColumn::MoveToZ(double z)        //沿 Z 轴平移
    {
        m_point.MoveToZ(z);
    }
    double C3DColumn::GetArea()const         //计算面积
    {
        double l = pi*2*m_r;
        double area = l*m_high + pi*m_r*m_r*2;
        return area;
    }
    double C3DColumn::GetVolume()const       //计算体积
    {
        double s_area = pi*m_r*m_r;
        return s_area*m_high;
    }
    void C3DColumn::show()const
    {
        m_point.show();
        cout << "---->:"<<m_r<<"== == == == > "<<m_high;
    }
    //================================================================
```

8.4 系统运行结果

针对上面的图形管理类层次结构模块，定义如下的测试程序。程序中通过一个函数创建了 20 个图形对象，然后在 main 函数中使用统一的接口显示这些图形对象的信息。创建图形对象的函数可以替换成从文件中读取图形信息等操作。在此这样处理的目的只是演示使用抽象类做界面程序，使得类定义模块与应用程序完全分离，成为一个完全独立的程序模块。这样便于软件代码的重用和形成软件构件，以及提高软件开发的效率。

```
    /****************************************************************
        File name   :  shape_driver.cpp
```

        Description：图形继承结构应用程序文件
***/
#include <iostream>
#include <string>
#include "shape.h"
#include "1DPoint.h"
#include "1DLine.h"
#include "2DPoint.h"
#include "2DCircle.h"
#include "2DLine.h"
#include "2DRectangle.h"
#include "2DTriangle.h"
#include "3DPoint.h"
#include "3DCube.h"
#include "3DColumn.h"
using namespace std;

typedef struct
{
    string name;
    CShape*shape;
}TagShape;

const int ShapeNum = 10;
/*函数用于构造一个包含 20 个图形的链表，以测试图形类库结构的正确性 ---------*/
int ReadShape(TagShape* shapList)
{
    int i = 0;
    int count = 0;

    for (i = 0; i < 20; i++)
    {
        try
        {
            int k = i % ShapeNum;
            switch (k){
            case 0:
                shapList[count].name = "C1DPoint";
                shapList[count].shape = new C1DPoint(i);
                break;
            case 1:
                shapList[count].name = "C1DLine";
                shapList[count].shape = new C1DLine(i, i+k);
                break;
            case 2:
                shapList[count].name = "C2DPoint";
                shapList[count].shape = new C2DPoint(i, i / ShapeNum+k);
```
```

```cpp
                    break;
                case 3:
                    shapList[count].name = "C2DLine";
                    shapList[count].shape = new C2DLine(i, i / ShapeNum + k,
                                       i + k, ShapeNum + k);
                    break;
                case 4:
                    shapList[count].name = "C2DRectangle";
                    shapList[count].shape = new C2DRectangle(i, i / ShapeNum + k,
                                       i + k, ShapeNum + k);
                    break;
                case 5:
                    shapList[count].name = "C2DTriangle";
                    shapList[count].shape = new C2DTriangle(0, 0, i, i, k, k + i);
                    break;
                case 6:
                    shapList[count].name = "C2DCircle";
                    shapList[count].shape = new C2DCircle(i, i, k + i / ShapeNum);
                    break;
                case 7:
                    shapList[count].name = "C3DPoint";
                    shapList[count].shape = new C3DPoint(i, i, i);
                    break;
                case 8:
                    shapList[count].name = "C3DCube";
                    shapList[count].shape = new C3DCube(i, i, i, k + i, k + i, k + i);
                    break;
                case 9:
                    shapList[count].name = "C3DColumn";
                    shapList[count].shape = new C3DColumn(i, i, i, k, k + i);
                    break;
                }
                ++count;
            }
        catch (string s)
        {
            cout << "Create shape is error: " << s << endl;
        }
    }
    return count;
}

int main()
{
    TagShape shapList[30];
    int count=0;
```

```
        count = ReadShape(shapList);
        for (int i = 0; i < count; i++)
        {
            cout << shapList[i].name << ": ";
            shapList[i].shape->show();
            cout << "\n";
        }

        return 0;
    }
```

上面应用程序的输出结果：

```
C1DPoint:    0
C1DLine:     1---->2
C2DPoint:    (2, 2)
C2DLine:     (3, 3)--->(6, 13)
C2DRectangle: (4, 4)=====>(8, 14)
C2DTriangle: (0, 0)---->(5, 5)---->(5, 10)---->(0, 0)
C2DCircle:   (6, 6)-----6
C3DPoint:    (7, 7, 7)
C3DCube:     (8, 8, 8)=======>(16, 16, 16)
C3DColumn:   (9, 9, 9)---->:9== == == == > 18
C1DPoint:    10
C1DLine:     11---->12
C2DPoint:    (12, 3)
C2DLine:     (13, 4)--->(16, 13)
C2DRectangle: (14, 5)=====>(18, 14)
C2DTriangle: (0, 0)---->(15, 15)---->(5, 20)---->(0, 0)
C2DCircle:   (16, 16)-----7
C3DPoint:    (17, 17, 17)
C3DCube:     (18, 18, 18)=======>(26, 26, 26)
C3DColumn:   (19, 19, 19)---->:9== == == == > 28
请按任意键继续. . .
```

练　习

1．使用面向对象程序设计的方法设计一个简单的图书管理系统，系统功能包括：
(1)普通图书：入库、借书、还书，借书周期3个月、报废处理。
(2)杂志管理：入库、借阅(不允许借出图书馆)。
(3)特别书籍：入库、借阅(不允许借出图书馆，但可以复印)。
(4)人员：图书管理员、读者。
2．使用面向对象程序设计的方法设计一个简单的学生学籍管理系统，系统功能包括：
(1)学生管理：增加学生、删除学生、显示学生信息、修改学生信息、查询学生信息。
(2)学生成绩管理功能：成绩录入、成绩输出、成绩统计、查询成绩等。

第 9 章 实 验

实验 1 指针、引用和函数重载

实验内容

(1) 用指针和引用作为函数编写如下函数：
a. 定义交换两个 int 变量的函数。

```
void swap(int *, int *);          //指针作为函数参数
void swap(int&, int&);            //引用作为函数参数
```

b. 定义交换两个 double 变量的函数。

```
void swap(double*, double*);      //指针作为函数参数
void swap(double&, double&);      //引用作为函数参数
```

c. 定义交换两个 string 类变量的函数。

```
void swap(string*, string*);      //指针作为函数参数
void swap(string&, string&);      //引用作为函数参数
```

d. 编写以上函数的测试程序。

(2) 用指针作为函数参数，编写如下函数：
a. 编写一个函数对 n 个 int 变量按照从小到大方式排序。

```
int* sort(int*, int);
```

b. 编写一个函数对 n 个 double 变量按照从小到大方式排序。

```
double* sort(double*, int);
```

c. 编写一个函数对 n 个 string 类对象按照从小到大方式排序。

```
string* sort(string*, int);
```

d. 编写以上函数的测试程序。

实验准备和说明

(1) 在学习完教材 1.1~1.2 节内容之后进行本次实验。
(2) 编写本次上机所需的程序并测试函数的正确性。
(3) 实验完毕，认真思考上述函数的不同，写出实验报告。

思考与练习

(1) 思考使用指针和引用作为函数参数时，函数的实现形式和功能有什么差异？

(2) 思考函数重载为软件编程带来的好处是什么？程序调用重载函数时，编译系统如何确定调用的具体函数？

实验2 文件操作

实验内容

手工创建一个不超过 100 个整型数的文本文件，编写程序完成如下功能：

(1) 编写一个函数，读取文件中的所有数据。

(2) 从键盘输入一个任意整数，编写一个函数判断(1)读取的整数中是否包含该整数，如果包含则返回包含的个数，否则返回-1。

实验准备和说明

(1) 在学习完教材 1.3 节内容之后进行本次实验。

(2) 编写本次上机所需的程序并测试函数的正确性。

(3) 实验完毕，认真思考文件操作的原理，写出实验报告。

思考与练习

(1) 思考文件的操作方式？

(2) 思考读取文件数据的不同方式的区别？

实验3 类的建立和使用

实验内容

给定一个中英文词典，试编写程序完成如下功能：

(1) 设计一个词典类，每个词条包括英文单词及其中文翻译。该类应该至少具有英汉对译的功能，即给出中文单词，能翻译成对应的英文单词；给出英文单词，能翻译成中文单词。

(2) 编写一个函数，读取文件中的词典数据。

(3) 给出一个英文句子，将其翻译成中文。

(4) 给出一个经过词语切分的中文句子，将其翻译成英文。

实验准备和说明

(1) 在学习完教材第 2、3 章内容之后进行本次实验。

(2) 编写本次上机所需的程序并测试函数的正确性。

(3) 实验完毕，写出实验报告。

思考与练习

(1) 中文翻译成英文时，存在哪些翻译问题，如何解决？
(2) 英文翻译成中文时，存在哪些翻译问题，如何解决？

实验 4 运算符重载

实验内容

定义一个分数类，该类具有分子和分母两个属性成员，试编写程序完成如下操作：
(1) 定义合适的构造函数。
(2) 定义前自增和后自增运算符重载函数，完成分子加 1 的操作。
(3) 定义两个分数的加法运算符重载函数。
(4) 定义两个分数的减法运算符重载函数。
(5) 定义两个分数的除法运算符重载函数。
(6) 定义两个分数的比较运算符重载函数。
(7) 定义流插入运算符重载函数，实现将分数显示在屏幕上。
(8) 定义流提取运算符重载函数，实现从键盘输入分数类对象的值。

实验准备和说明

(1) 在学习完教材第 4 章内容之后进行本次实验。
(2) 编写本次上机所需的程序并测试函数的正确性。
(3) 实验完毕，认真思考运算符重载的原理，写出实验报告。

思考与练习

思考运算符重载成员函数和友元函数的不同？

实验 5 继承与派生

实验内容

分别定义 CTeacher(教师)类和 CAdministrator(管理干部)类，采用多重继承方式，由这两个类派生出新类 CTeacher_Administrator(教师兼管理干部)。要求：
(1) 在两个基类中都包含姓名、性别、地址、电话等数据成员。
(2) 在 CTeacher 类中还包含数据成员 title(职称)，在 CAdministrator 类中还包含数据成员 post(职务)，在 CTeacher_Administrator 类中还包含数据成员 wages(工资)。
(3) 对两个基类中的姓名、性别、地址、电话等数据成员用相同的名字，在引用这些数据成员时，指定作用域。
(4) 在类体中声明成员函数，在类外定义成员函数。
(5) 在派生类 CTeacher_Administrator 的成员函数 show 中调用 Teacher 类中的 display 函数，输出姓名、性别、职称、地址、电话，然后再用 cout 语句输出职务与工资。

实验准备和说明

(1) 在学习完教材的第 5 章继承和派生之后进行本实验。
(2) 编写本次上机所需要的程序。

实验步骤

(1) 打开 VisualStudio，并创建 C++新项目 Exam5。
(2) 设计 CTeacher 类，并在 exam.cpp 中输入下列代码：

```cpp
#include <string>
#include <iostream>
using namespace std;
class CTeacher
{
public:
    CTeacher(string nam,string s,string tit,string ad,string t);
    void display();
protected:
    string name; string sex; string title; string address; string tel;
};
CTeacher::CTeacher(string nam,string s,string tit,string ad,string t)
        : name(nam),sex(s),title(tit),address(ad),tel(t)
{ }
void CTeacher::display()
{
    cout<<"姓名:"<<name<<endl;
    cout<<"性别:"<<sex<<endl;
    cout<<"职称:"<<title<<endl;
    cout<<"地址:"<<address<<endl;
    cout<<"电话:"<<tel<<endl;
}
```

(3) 设计并添加 CAdministrator 类代码：

```cpp
CAdministrator::CAdministrator(string nam,string s,string p,string
        ad,string t): name(nam),sex(s),post(p),address(ad),tel(t)
{}
void CAdministrator::display()
{
    cout<<"name:"<<name<<endl;
    cout<<"sex:"<<sex<<endl;
    cout<<"post:"<<post<<endl;
    cout<<"address:"<<address<<endl;
    cout<<"tel:"<<tel<<endl;
}
```

(4) 设计并添加 CTeacher_Administrator 类代码：

```cpp
CTeacher_Administrator::CTeacher_Administrator(string nam,string s,string t,
```

```
            string p,string ad,string tel,float w): CTeacher(nam,s,t,ad,tel),
            CAdministrator(nam,s,p,ad,tel),wage(w)
    {}
    void CTeacher_Administrator::show()
    {
        CTeacher::display();
        cout<<"职务:"<<CAdministrator::post<<endl;
        cout<<"工资:"<<wage<<endl;
    }
```

(5) 添加 main 函数及测试代码：

```
    int main()
    {
        CTeacher_Administrator a("张英","男","高级","教师","宁海路122号计算中心",
                              "13800000000",5000);
        a.show();
        system("pause");
        return 0;
    }
```

编译运行并测试，写出运行结果。

(6) 写出实验报告。

实验 6 多态性和虚函数

实验内容

在银行账户中，一个基础账号 Account 类，包含账号和余额信息，作为基类。一个从基础账号 Account 类派生的存款账号 Savings 类，包含最小余额信息。另一个从基础账号 Account 类派生的结算账号类 Checking 类，包含汇款方式信息。

AccountList 类定义了一个链表，通过链表可以将银行的各个账户对象组织在这个双向链表中，从而实现了类似容器的功能。在链表中通过指针来操作元素，可以体现出元素的多态性。

程序采用多文件结构，包含主程序文件在内一共 9 个文件，分别是：

基础账号类：

- account.h
- account.cpp

存款账号类：

- savings.h
- savings.cpp

结算账号类：

- checking.h
- checking.cpp

链表容器类：

- accountlist.h
- accountlist.cpp

主程序:
- linked_list.cpp

实验准备和说明

(1)在学习完教材的第 6 章多态性和虚函数之后进行本实验。
(2)编写本次上机所需要的程序。

实验步骤

(1)打开 VisualStudio,并创建 C++新项目 Exam6。
(2)设计 Account 类,在新建文件 account.h 和 account.cpp 中输入下列代码:

```
/*********************************************************************
    File name   : account.h
    Description : account.h 部分
**********************************************************************/
#ifndef HEADER_ACCOUNT
#define HEADER_ACCOUNT
//------------------------------------
#include<string>
using std::string;
//------------------------------------
class Account
{
protected:
    string acntNumber;                          //账号
    double balance;                             //余额
public:
    Account(string acntNo, double balan=0.0);
    virtual void display()const;
    double getBalan()const
{
return balance;
}
    void deposit(double amount)
{
balance += amount;
}
    bool operator==(const Account& a)
{
return acntNumber==a.acntNumber;
}
    string getAcntNo()
{
return acntNumber;
```

```
        virtual void withdrawal(double amount)=0;
};
//------------------------------------
#endif                                    //HEADER_ACCOUNT


/*********************************************************************
File name   : account.cpp
Description : account.cpp 部分
*********************************************************************/
#include<iostream>
#include"account.h"
//------------------------------------
Account::Account(string acntNo, double balan)
:acntNumber(acntNo),balance(balan)
{}
//------------------------------------

void Account::display()const
{
std::cout<<"Account:"+acntNumber+" = "<<balance<<"\n";
}
//------------------------------------
```

(3) 设计 Account 类的派生类 Savings 类，在新建文件 savings.h 和 savings.cpp 中输入下列代码：

```
/*********************************************************************
File name   : savings.h
Description : savings.h 部分
*********************************************************************/
#ifndef HEADER_SAVINGS
#define HEADER_SAVINGS
//------------------------------------
#include"account.h"
#include<string>
using std::string;
//------------------------------------
class Savings : public Account
{
public:
    Savings(string acntNo, double balan=0.0):Account(acntNo,balan)
    {}
    void display()const;
    void withdrawal(double amount);
};
//------------------------------------
#endif                                    //HEADER_SAVINGS
```

```
/******************************************************************
File name    : savings.cpp
Description  : savings.cpp 部分
******************************************************************/
#include<iostream>
#include"savings.h"
//-------------------------------------
void Savings::display()const
{
    std::cout<<"Savings ";
    Account::display();
}
//-------------------------------------
void Savings::withdrawal(double amount)
{
    if(balance < amount)
        std::cout <<"Insufficient funds withdrawal: "<<amount<<"\n";
    else
        balance -= amount;
}
//-------------------------------------
```

(4) 设计 Account 类的派生类 Checking 类，在新建文件 Checking.h 和 Checking.cpp 中输入下列代码：

```
/******************************************************************
File name    : checking.h
Description  : checking.h 部分
******************************************************************/
#ifndef HEADER_CHECKING
#define HEADER_CHECKING
#include"account.h"
//-------------------------------------
enum REMIT{remitByPost, remitByCable, other};   //信汇, 电汇, 无
//-------------------------------------
class Checking : public Account
{
    REMIT remittance;                           //汇款方式
public:
    Checking(string acntNo, double balan=0.0);
    void withdrawal(double amount);
    void display()const;
    void setRemit(REMIT re)
{
remittance = re;
}
};
//-------------------------------------
#endif                                          //HEADER_CHECKING
```

```
/******************************************************************
File name   : checking.cpp
Description : checking.cpp 部分
*******************************************************************/
#include<iostream>
#include"checking.h"
Checking::Checking(string acntNo, double balan)
:Account(acntNo, balan), remittance(other)
{
}
//----------------------------------
void Checking::display()const
{
    std::cout<<"Checking ";
    Account::display();
}
//----------------------------------
void Checking::withdrawal(double amount)
{
    if(remittance==remitByPost)                //信汇30元手续费
    amount += 30;
    if(remittance==remitByCable)               //电汇60元手续费
    amount += 60;
    if(balance < amount)
    std::cout <<"Insufficient funds withdrawal: "<<amount<<"\n";
    else
    balance -= amount;
}
```

(5) 设计 AccountList 类和 Node 类，在新建文件 accountlist.h 和 accountlist.cpp 中输入下列代码：

```
/******************************************************************
File name   : accountlist.h
Description : accountlist.h 部分
*******************************************************************/
#ifndef ACCOUNTLIST
#define ACCOUNTLIST
#include"account.h"
//-------------------------------------
class Node
{
public:
    Account& acnt;
    Node *next, *prev;                     //双向链表，每个结点前后指针
    Node(Account& a):acnt(a),next(0),prev(0)
    {}
    bool operator==(const Node& n)const
```

```cpp
    {
        return acnt==n.acnt;
    }
};
//----------------------------------

class AccountList
{
    int size;
    Node *first;
public:
    AccountList():first(0),size(0)
    {}
    Node* getFirst()const
    {
        return first;
    }
    int getSize()const
    {
        return size;
    }
    void add(Account& a);
    void remove(string acntNo);
    Account* find(string acntNo)const;
    bool isEmpty()const
    {
        return !size;
    }
    void display()const;
    ~AccountList();
};
#endif                                                  //HEADER_ACCOUNTLIST

/**********************************************************************
File name   : accountlist.cpp
Description : accountlist.cpp 部分
**********************************************************************/
#include<iostream>
using std::cout;
#include"accountlist.h"
//----------------------------------------
void AccountList::add(Account& a)
{
    Node* pN = new Node(a);
    if(first)
    {
        pN->next = first;
        first->prev = pN;
```

```
    }
    first = pN;
    size++;
}
//----------------------------------
void AccountList::remove(string acntNo)
{
    for(Node* p=first; p; p=p->next)
    if((p->acnt).getAcntNo()==acntNo)
    {
        if(p->prev)p->prev->next = p->next;
        if(p->next)p->next->prev = p->prev;
        if(p==first)first = p->next;
        delete p;
        size--;
        break;
    }
}
//----------------------------------
Account* AccountList::find(string acntNo)const
{
    for(Node* p=first; p; p=p->next)
        if((p->acnt).getAcntNo()==acntNo)
            return &(p->acnt);
    return 0;
}
//----------------------------------
void AccountList::display()const
{
    std::cout<<"There are "<<size<<" accounts.\n";
    for(Node* p=first; p; p=p->next)
        (p->acnt).display();
}
//----------------------------------
AccountList::~AccountList()
{
    for(Node* p=first; p=first; delete p)
        first = first->next;
}
//----------------------------------
```

(6) 添加 main 函数及测试代码：

```
/*****************************************************************
File name  : linked_list.cpp
Description : 主程序部分
*****************************************************************/
#include<iostream>
using namespace std;
```

```cpp
#include"savings.h"
#include"checking.h"
#include"accountlist.h"
int main()
{
    Savings s1("101",2000), s2("102", 4000);
    Checking c1("201"), c2("312", 10000);
    AccountList a;
    a.add(s1);
    a.add(s2);
    a.add(c1);
    a.add(c2);
    Account* p;
    if(p = a.find("101"))p->deposit(100);
    if(p = a.find("201")) p->deposit(2000);
    if(p = a.find("102"))p->withdrawal(2500);
    if(p = a.find("312")) p->withdrawal(2950);
    a.display();
    system("pause");
    return 0;
}
```

编译运行并测试，写出运行结果。

(7) 写出实验报告。

实验7　文件系统的操作

实验内容

(1) 给定《人民日报》1998年1月1日到10日的语料，分别存于文件 rmrb19980101.txt 到 rmrb19980110.txt 中，并存在 C:/rmrb/文件夹下。

(2) 编写程序，对10日人民日报中的语料统计其中二元组的频度，并按照频度降序排列，存入 bigram.txt 文件中。

实验准备和说明

(1) 在学习完第7章后进行本实验。
(2) 编写本次上机所需的程序并测试函数的正确性。
(3) 实验完毕，阅读 bigram.txt 文件，分析结果是否正确，写出实验报告。

思考与练习

(1) 程序编写过程中遇到了哪些困难？出现了哪些错误？如何解决？
(2) 统计二元组与统计单词的词频在程序设计上有哪些不同？要考虑哪些问题？

参 考 文 献

陈波，吉根林．2010．C语言程序设计教程．北京：中国铁道出版社．
陈卫卫．2002．C/C++程序设计教程．北京：希望电子出版社．
陈小荷．2000．现代汉语自动分析：Visual C++实现．北京：北京语言大学出版社．
吉根林，陈波．2014．数据结构教程（C++语言描述）．北京：高等教育出版社．
吕凤翥．2007．C++语言程序设计．2版．北京：电子工业出版社．
罗建军，朱丹军，顾刚，等．2004．C++程序设计教程．北京：高等教育出版社．
钱能．2014．C++程序设计教程详解——过程化编程．北京：清华大学出版社．
孙淑霞，肖阳春，魏琴．2008．C/C++程序设计教程．2版．北京：电子工业出版社．
谭浩强．2011．C++程序设计．2版．北京：清华大学出版社．
吴乃陵，况迎辉，李海文．2003．C++程序设计．北京：高等教育出版社．
徐孝凯．2011．C++语言基础教程．2版．北京：清华大学出版社．
余苏宁，王明福．2003．C++程序设计．北京：高等教育出版社．
赵建周，杨庆祥．2000．C和C++程序设计教程．北京：航空工业出版社．
郑阿奇，丁有和．2009．C++面向对象实用教程．北京：电子工业出版社．
郑莉，董渊，张瑞丰．2014．C++语言程序设计．3版．北京：清华大学出版社．
Carrano F M，Helman P，Veroff R．1998．Data Abstraction and Problem Solving with C++．Addison Wesley．
Deitel H M，Deitel P J．2004．C++程序设计教程．4版．施平安，译．北京：清华大学出版社．
Lafore R．1998．Objected-Oriented Programming in C++．3rd ed．SAMS．
Lippman S B，LajoieJ．2013．C++ Primer（英文版）．5版．北京：电子工业出版社．

附 录

模拟试卷 1

一、选择题(每题 2 分,共 20 分)

1. 面向对象程序设计思想的主要特征中不包括(   )。
   A. 封装性　　　B. 多态性　　　C. 继承性　　　D. 功能分解,逐步求精
2. 关于类和对象不正确的说法是(   )
   A. 类是一种类型,它封装了数据和操作　　　B. 对象是类的实例
   C. 一个类的对象只有一个　　　D. 一个对象必属于某个类
3. 一个类的友元函数能够访问该类的(   )。
   A. 私有成员　　　B. 保护成员　　　C. 公有成员　　　D. 所有成员
4. 不能作为函数重载判断依据的是(   )。
   A. 参数个数　　　B. 参数类型　　　C. 函数名字　　　D. 返回类型
5. 下面对构造函数的不正确描述是(   )。
   A. 系统可以提供默认的构造函数
   B. 构造函数可以有参数,所以也可以有返回值
   C. 构造函数可以重载
   D. 构造函数可以设置默认参数
6. 下述静态数据成员的特征中,(   )是错误的。
   A. 说明静态数据成员时前边要加修饰符 static
   B. 静态数据成员要在类体外进行初始化
   C. 引用静态数据成员时,要在静态数据成员名前加<类名>和作用域运算符
   D. 静态数据成员不是该类所有对象所共用的
7. (   )是析构函数的特征。
   A. 一个类中只能定义一个析构函数　　　B. 析构函数名与类名无关
   C. 析构函数的定义只能在类体内　　　D. 析构函数可以有 1 个或多个参数
8. 关于 new 运算符的下列描述中,(   )是错误的。
   A. 它可以用来动态创建对象和对象数组
   B. 使用它创建对象或对象数组,可以使用运算符 DELETE 删除
   C. 使用它创建对象时要调用构造函数
   D. 使用它调用对象数组时不许指定初始值
9. 在重载一个运算符时,如果其参数表中有一个参数,则说明该运算符是(   )。
   A. 一元成员运算符　　　B. 二元成员运算符
   C. 一元友元运算符　　　D. 选项 B 和 C 都可能

10. 下列对派生类的描述中，错误的是（   ）。
    A. 一个派生类可以作为另一个派生类的基类
    B. 派生类至少有一个基类
    C. 派生类的成员除了自己的成员外，还包含它的基类的成员
    D. 派生类中继承的基类成员的访问权限到派生类保持不变

二、阅读程序并写出运行结果（每题5分，共25分）

1. 程序如下：

```cpp
#include<iostream.h>
class Sample
{
    int i;
    double d;
    public:
        void setdata(int n){i=n;}
        void setdata(double x){d=x;}
        void disp()
        {
            cout<<"i="<<i<<",d="<<d<<endl;
        }
};
void main()
{
    Sample s;
    s.setdata(10);
    s.setdata(15.6);
    s.disp();
}
```

2. 分析下面的程序，写出其运行时的输出结果。

```cpp
#include <iostream.h>
class A
 {
public:
    A(){cout<<"1";}
    ~A(){cout<<"2";}
 };
class B:public A
{
    public:
      B(){cout<<"3";}
     ~B(){cout<<"4";}
};
void main()
{
    B b;
}
```

3. 分析以下程序的执行结果。

```cpp
#include "stdafx.h"
#include<iostream>
using namespace std;
class T
{ public:
    T( int x){ a=x; b+=x;};
    static void display( T c)
    {cout<<"a="<<c.a<<'\t'<<"b="<<c.b<<endl;}
 private:
    int a;
    static int b;
};
int T::b=5;
void main()
{ T A(3),B(5);
 T::display(A);
 T::display(B);
}
```

4. 分析以下程序的执行结果。

```cpp
#include<iostream.h>
class Sample
{
    int x;
    public:
        Sample(){};
        Sample(int a){x=a;}
        Sample(Sample &a){x=a.x+1;}
        void disp(){cout<<"x="<<x<<endl;}
};
void main()
{
    Sample s1(2),s2(s1);
    s2.disp();
}
```

5. 分析下面的程序，写出其运行时的输出结果。

```cpp
#include <iostream>
using namespace std;
class Base
{
  public:
    virtual void f(){ cout << "f0+"; }
    void g(){ cout << "g0+"; }
};
class Derived : public Base
```

```
{
 public:
    void f(){ cout << "f+"; }
    void g(){ cout << "g+"; }
};
void main(){ Derived d;  Base *p = &d;  p->f();  p->g(); }
```

三、完善程序（共 25 分）

1. 写一个函数 length，求一个字符串的长度，在 main 函数中输入字符串，并输出其长度，请填空。（10 分）

```
void main()
{
    int len;
    char *str[20];
    printf("please input a string:\n");
    scanf("%s",str);
    len=length(str);
    printf("the string has %d characters.",len);
}
int length(char *p)
{
    int n;
    n=0;
    while(*p!='\0')
    {
        ___【1】___ ;
        ___【2】___ ;
    }
    return n;
}
```

2. 以下程序是定义一个计数器类 counter，对其重载运算符"+"，请填空。（5 分）

```
class counter
{ private: int n;
  public:
    counter(){n=0;}
    counter(int i){n=i;}
    ___【3】___   //运算符重载函数
    { counter t; t.n=n+c.n; return t; }
    void disp(){cout<<"n="<<n<<endl;}
};
void main()
{ counter c1(5),c2(10),c3;
  c3=c1+c2;
  c1.disp(); c2.disp(); c3.disp(); }
```

3. 下列程序的输出结果为 2,请将程序补充完整。(10 分)

```cpp
class Base
{ public:
    ___【4】___ void fun()
    { cout<<1; }
};
class Derived: public Base
{ public:
    void fun(){ cout<<2; }
};
int main()
{ Base *p= new Derived;
  p->fun();
    ___【5】___ ;//删除对象
  return 0;
}
```

四、编程题(共 30 分)

1. 根据下面给定的代码,完成点的位移的+、-操作。(10 分)

```cpp
#include<iostream>
using namespace std;
class point
{
  int x,y;
  public:
   point (int vx,int vy) {x=vx; y=vy;}
   point() {x=0; y=0;}
   point operator +(point p1);
   point operator-(point p1);
   void print(){cout<<x<<"  "<<y<<"\n";}
};

void main()
{
  point p1(50,40), p2(30,10);
  p1=p1 - p2;
  p1.print();
}

point point::operator + (point p1)
{
    ……
}

point point::operator - (point p1)
{
    ……
}
```

2. 建立一个类名为 array 的类，类中存储一个 4×4 的矩阵并能查找其中最大的数。例如，以下矩阵中的最大数是 16，它在第 2 行、第 2 列。(10 分)

```
  5   6   7   8
 10   9  12  11
 14  13  16  15
  4   3   2   1
```

类定义的要求如下：

(1) 私有数据成员：

　　int p[4][4] ——存储一个 4×4 的矩阵；

　　int n ——矩阵的行列数。

(2) 公有数据成员：

　　int x 　　——存储要查找的最大数；

　　int im,jm ——要查找的最大数在矩阵中的行和列。

(3) 公有函数成员：

　　构造函数——初始化 n 的值为 4，x,im,jm 为 0。

　　void input(int a[][4])——将矩阵赋给该成员中的数组。

　　void find()——在该函数中查找矩阵中的最大值，并把最大值及其所在行列保存到 x,im,jm 中。

　　void print ——按行输出矩阵值。

其中主函数为：

```
void main(){
    int a[][4]={5,6,7,8,10,9,12,11,14,13,16,15,4,3,2,1};
    array s;
    s.input(a);
    s.find();
    s.print();
}
```

3. 设计程序，有基类 Base，从基类派生圆类 Circular，再从圆类 Circular 派生圆柱类 Column，设计成员函数输出它们的面积和体积。(10 分)

其中基类定义为：

```
class Basic                    //基类
{
    public:
        virtual double getArea()=0;
};
```

模拟试卷 2

一、选择题(每题 2 分，共 20 分)

1. 下面概念中，不属于面向对象方法的是(　　)。

A. 对象、消息　　B. 继承、多态　　C. 类、封装　　D. 过程调用
2. 在类定义对象的外部，可以被访问的成员有（　　）。
   A. 所有类成员　　　　　　　　B. private 或 protected 的类成员
   C. public 的类成员　　　　　　D. public 或 private 的类成员
3. 静态成员函数没有（　　）。
   A. 返回值　　B. this 指针　　C. 指针参数　　D. 返回类型
4. 在 C++语言中，对函数参数默认值描述正确的是（　　）。
   A. 函数参数不能设定默认值
   B. 一个函数的参数若有多个，则参数默认值的设定可以不连续
   C. 函数参数必须设定默认值
   D. 在设定了某参数的默认值后，该参数后面定义的所有参数都必须设定默认值
5. 下面（　　）项是对构造函数和析构函数的正确定义。
   A. void X::X(), void X::~X()　　　　B. X::X(参数), X::~X()
   C. X::X(参数), X::~X(参数)　　　　D. void X::X(参数), void X::~X(参数)
6. 复制构造函数具有的下列特点中，（　　）是错误的。
   A. 如果一个类中没有定义复制构造函数时，系统将自动生成一个默认的
   B. 复制构造函数只有一个参数，并且是该类对象的引用
   C. 复制构造函数是一种成员函数
   D. 复制构造函数的名字不能用类名
7. 假定 MyClass 为一个类，则执行 MyClass a[3],*p[2];语句时，自动调用该类构造函数（　　）次。
   A. 2　　　　B. 3　　　　C. 4　　　　D. 5
8. 下列对派生类的描述中，（　　）是错误的。
   A. 一个派生类可以作为另一个派生类的基类
   B. 派生类至少应有一个基类
   C. 基类中成员访问权限继承到派生类中都保持不变
   D. 派生类的成员除了自己定义的成员外，还包含了它的基类成员
9. 关于下列虚函数的描述中，（　　）是正确的。
   A. 虚函数是一个 static 存储类的成员函数
   B. 虚函数是一个非成员函数
   C. 基类中说明了虚函数后，派生类中其对应的函数可不必说明为虚函数
   D. 派生类的虚函数与基类的虚函数应具有不同的类型或个数
10. 下列描述中，（　　）是抽象类的特征。
    A. 可以说明虚函数　　　　　　B. 可以进行构造函数重载
    C. 可以定义友元函数　　　　　D. 不能定义其对象

二、阅读程序并写出运行结果(每题 5 分，共 25 分)

1. 写出如下程序的运行结果。

```
#include<iostream>
using namespace std;
```

```cpp
class Test{
public:
    Test(){}
    ~Test(){cout<<'#';}
};
int main(){
    Test temp[2], *pTemp[2];
    return 0;
}
```

2. 分析下面的程序，写出其运行时的输出结果。

```cpp
#include<iostream.h>

class point
{
  int x,y;
  public:
    point (int vx,int vy) {x=vx; y=vy;}
    point()   {x=0; y=0;}
    point operator +(point p1);
    point operator-(point p1);
    void print(){cout<<x<<"  "<<y<<"\n";}
};

point point::operator + (point p1)
{
    point p;
    p.x=x + p1.x;
    p.y=y + p1.y;
    return p;
}

point point::operator - (point p1)
{
    point p;
    p.x=x - p1.x;
    p.y=y - p1.y;
    return p;
}

void main()
{
  point p1(50,40), p2(30,10);
  p1=p1 - p2;
  p1.print();
}
```

3. 写出程序的运行结果。

```cpp
#include <iostream.h>
class B1
{
public:
    int nv;
    void fun(){cout<<"Member of B1."<<endl;}
};
class B2
{
public:
    int nv;
    void fun(){cout<<"Member of B2."<<endl;}
};
class D1:public B1,public B2
{
public:
    int nv;
    void fun(){cout<<"Member of D1."<<endl;}
};

void main()
{
    D1 d1;
    d1.nv=1;
    d1.fun();
    d1.B1::nv=2;
    d1.B1::fun();
    d1.B2::nv=3;
    d1.B2::fun();
}
```

4. 分析下面的程序，写出其运行时的输出结果。
有如下程序：

```cpp
#include<iostream.h>
class BASE{
    char c;
public:
    BASE(char n):c(n){}
    ~BASE(){cout<<C;}
};
class DERIVED:public BASE{
    char c;
public:
    DERIVED(char n):BASE(n+1),c(n){}
    ~DERIVED(){cout<<C;}
};
```

```
int main()
{   DERIVED("X");
    return 0;
}
```

5. 分析下列程序的输出结果。

```
#include<iostream>
using namespace std;
    class B
{
public:
    B(){}
    B(int i){b=i;}
    virtual void virfun()
    {
        cout<<"B::virfun()called.\n";
    }
private:
    int b;
};
class D:public B
{
public:
    D(){}
    D(int i,int j):B(i){d=j;}
private:
    int d;
    void virfun()
    {
        cout<<"D::virfun()called.\n";
    }
};
void fun(B *obj)
{
    obj->virfun();
}
int main()
{
    D *pd=new D;
    fun(pd);
    return 0;
}
```

三、完善程序(共 25 分)

1. 下面是字符串复制程序，请填空。(10 分)

```
void ccopy()
{   char ch1[]="good morning!",ch2[20];
```

```
        int i=0;
        while(   【1】   )
        {   ch2[i]=ch1[i];
            i++;
        }
        【2】
}
```

2. 填空。(5分)

```
#include <iostream>
using namespace std;
class Base
{ public:
    void fun(){cout<<"Base::fun"<<endl;}
};
class Derived:public Base
{   void fun()
    {  【3】    //显示调用基类的函数 fun()
        cout<<"Derived::fun"<<endl;
    }
};
```

3. 请将下面的程序补充完整，使得程序输出"平凡的世界是路遥的书"。(10分)

```
#include <iostream>
#include"string.h"
using namespace std;
class Book{
public:
    Book(char *str){strcpy(title,str);}
    【4】   void PrintInfo(){cout<<title<<endl;}
protected:
    char title[50];
};
class MyBook:public Book{
public:
    MyBook(char *s1,char *s2="路遥"):  【5】
    {strcpy(owner,s2);}
    virtual void PrintInfo(){cout<<title<<"是"<<owner<<"的书"<<endl;}
private:
    char owner[10];
};
int main(){
    Book *prt=new MyBook("平凡的世界");
    prt->PrintInfo();
    return 0;
}
```

四、编程题(共 30 分)

1. 定义一个日期类 Date，包含年、月、日三个数据成员，以及一个求第二天日期的成员函数和输出日期的成员函数。(10 分)

注：
(1) 判断闰年的方法(year%400==0||year%4==0&&year%100!=0)为真 year 为闰年；
(2) 从 1~12 月，每个月的天数，分闰月和不闰月。

```
totaldays[2][12]={{31,28,31,30,31,30,31,31,30,31,30,31},{31,29,31,30,
                  31,30,31,31,30,31,30,31}};
```

2. 定义一个 Dog 类，它用静态数据成员 Dogs 记录 Dog 的个体数目。每创建一个对象 Dog 加 1，每释放一个对象 Dog 减 1。静态成员函数 GetDogs 用来读取 Dogs，SeeDogs 用来设定当前 Dog 的个数。设计并测试这个类。(10 分)

其中主函数为：

```
public:
    virtual float area( )=0;
};
float total(shape *s[ ],int n)
```

3. 下列 shape 类是一个表示形状的抽象类，area() 为求图形面积的函数，total() 则是一个通用的用以求不同形状的图形面积总和的函数。从 shape 类派生三角形类(triangle)、矩形类(rectangle)，并给出具体的求面积函数。只要求完成三角形类(triangle)、矩形类(rectangle) 的定义。(10 分)

```
class shape{
public:
    virtual float area()=0;
};
float total(shape *s[ ],int n)
{
    float sum=0.0;
    for(int i=0;i<n;i++)
    sum+=s[i]->area();
     return sum;
}
```

模拟试卷 3

一、选择题(每题 2 分，共 20 分)

1. 若已经声明了函数原型 "void fun(int a, double b=0.0);"，则下列重载函数声明中正确的是( )。

    A. void fun(int a=90, double b=0.0);
    B. int fun(int a, double B);
    C. void fun(double a, int B);
    D. bool fun(int a, double b = 0.0);

2. 如下函数定义：

```
void func(int a,int&b){a++; b++;}
```

若执行下面代码段后：

```
int x=0,y=1;
func(x,y);
```

则变量 x 和 y 值分别是（   ）。

  A．0 和 1   B．1 和 1   C．0 和 2   D．1 和 2

3．若 MyClass 是一个类名，且有如下语句序列：

```
MyClass c1,*c2;
MyClass *c3=new MyClass;
MyClass &c4=c1;
```

上面的语句序列所定义的类对象的个数是（   ）。

  A．1   B．2   C．3   D．4

4．派生类的成员函数不能访问基类的（   ）。

  A．有成员和保护成员   B．公有成员

  C．私有成员   D．保护成员

5．有以下程序：

```
#include<iostream>
using namespace std;
class MyClass
{
public:
    MyClass(int n){number = n;}
    //复制构造函数
    MyClass(MyClass &other){ number=other.number;}
    ~MyClass(){}
private:
    int number;
};
MyClass fun(MyClass p)
{
    MyClass temp(p);
    return temp;
}
int main()
{
    MyClass obj1(10), obj2(0);
    MyClass obj3(obj1);
    obj2=fun(obj3);
    return 0;
}
```

程序执行时，MyClass 类的复制构造函数被调用（   ）次。

  A．5   B．4   C．3   D．2

6．对于通过公有继承定义的派生类，若其成员函数可以直接访问基类的某个成员，说明该基类成员的访问权限是（   ）。

A. 公有或私有　　B. 私有　　　　C. 保护或私有　　D. 公有或保护

7. 有如下程序：

```
#include<iostream>
using namespace std;
class A
{
public:
    A(int i){ x= i;}
    void dispa(){ cout<<x<<',';}
private:
    int x;
};
class B:public A
{
public:
    B(int i):A(i +10){ x =i;}
    void dispb(){ dispa();cout<<x<< endl;}
private:
    int x;
};
int main()
{
    B b(2);
    b.dispb();
    return 0;
}
```

运行时输出的结果是（　　）。

A. 10，2　　　　B. 12，10　　　　C. 12，2　　　　D. 2，2

8. 下面是类 Shape 的定义：

```
class Shape{
    public:
    virtual void Draw()=0;
};
```

下列关于 Shape 类的描述中，正确的是（　　）。

A. 类 Shape 是虚基类

B. 类 Shape 是抽象类

C. 类 Shape 中的 Draw 函数声明有误

D. 语句"Shape s;"能够建立 Shape 的一个对象 s

9. 有如下程序：

```
#include <iostream>
using namespace std;
class B{
public:
```

```
    virtual void show(){cout<<"b";}
};
class D:public B {
public:
    void show(){cout<<"d";}
};
    void fun1(B*ptr){ptr->show();}
    void fun2(B &ref){ref.show();}
    void fun3(B b){b.show();}
    int main(){
       B b,*p=new D;
       D d;
       fun1(p);
       fun2(b);
       fun3(d);
       return 0;
    }
```

程序的输出结果是( )。

  A. ddb   B. bbd   C. dbb   D. dbd

10. 打开文件时可单独或组合使用下列文件打开模式：

①ios_base::in；②ios_base::binary；③ios_base::app；④ios_base::out。

若要以二进制添加方式打开一个文件，需使用的文件打开模式为( )。

  A. ①③   B. ①④   C. ②③   D. ②④

二、程序分析题(每小题 4 分，共 40 分)

1. 写出下面程序的输出结果。

```cpp
#include <iostream>
using namespace std;
class point
{ public:
    void poi(int px=10,int py=10)
    { x=px; y=py;}
    int getpx()
    { return x;}
    int getpy()
    { return y;}
  private:
    int x,y;
};
int main()
{ point p,q;
  p.poi();
  q.poi(15,15);
  cout<<"p 点的坐标是："<<p.getpx()<<", ";
  cout<<p.getpy()<<endl;
```

```
            cout<<"q 点的坐标是: "<<q.getpx()<<", ";
            cout<<q.getpy()<<endl;
            return 0;
        }
```

2. 写出下面程序的输出结果。

```
#include<iostream>
using namespace std;
//--------------------------------

class Point{
  int x, y;
public:
  void set(int a, int b){ x=a, y=b; }
  void print()const{ cout<<"("<<x<<", "<<y<<")\n"; }
  Point operator+(const Point& a);
  friend Point add(const Point& a, const Point& b);
};//===============================
Point Point::operator+(const Point& a){
  Point s;
  s.set(a.x+x, a.y+y);
  return s;
}//-------------------------------
Point add(const Point& a, const Point& b){
  Point s;
  s.set(a.x+b.x, a.y+b.y);
  return s;
}//-------------------------------
Void main(){
  Point a, b;
  a.set(3,2);
  b.set(1,5);
  (a+b).print();
  add(a, b).print();
}
```

3. 写出下面程序的输出结果。

```
#include <iostream>
using namespace std;
class Test {
public:
    Test(){ n+=2; }
    ~Test(){ n-=3; }
    static int getNum(){ return n; }
private:
    static int n;
};
int Test::n = 1;
```

```
int main(){
    Test* p = new Test;
    delete p;
    cout << "n=" << Test::getNum()<< endl;
    return 0;}
```

4. 写出下面程序的输出结果。

```
#include<iostream>
#include <string>
using namespace std;
//------------------------------------
class StudentID{
  int value;
public:
  StudentID(int id=0){
    value=id;
    cout <<"Assigning student id " <<value <<endl;
  }
};//------------------------------------
class Student{
  string name;
  StudentID id;
public:
  Student(string n="no name", int ssID=0):id(ssID),name(n){
    cout<<"Constructing student "<<n<<"\n";
  }
};//------------------------------------
Void main(){
  Student s("Randy", 98);
  Student t("Jenny");
}
```

5. 写出下面程序的输出结果。

```
#include<iostream>
using namespace std;
//------------------------------------
class A{
public:
  A(){ cout<<"A->"; }
  ~A(){ cout<<"<-~A"; }
};//------------------------------------
class B{
public:
  B(){ cout<<"B->"; }
  ~B(){ cout<<"<-~B"; }
};//------------------------------------
class C{
public:
```

```
    C(){ cout<<"C->"; }
    ~C(){ cout<<"<-~C"; }
};//----------------------------------
void func(){
    cout<<"\nfunc: ";
    A a;
    cout<<"ok->";
    static B b;
    C c;
}//------------------------------------
Void main(){
    cout<<"main:";
    func();
    func();
}
```

6. 写出下面程序的输出结果。

```
#include<iostream>
using namespace std;
//------------------------------------
class Person{
    char* pName;
public:
    Person(char* pN="noName"){
        cout<<"Constructing "<<pN<<"\n";
        pName = new char[strlen(pN)+1];
        if(pName)strcpy(pName,pN);
    }
    Person(const Person& s){
        cout<<"copy Constructing "<<s.pName<<"\n";
        pName = new char[strlen(s.pName)+1];
        if(pName)strcpy(pName, s.pName);
    }
    ~Person(){
        cout <<"Destructing "<<pName<<"\n";
        delete[] pName;
    }
};//---------------------------------
Void main(){
    Person p1("Randy");
    Person p2(p1);
}
```

7. 写出下面程序的输出结果。

```
#include <iostream>
using namespace std;
class A{
    int x;
```

```cpp
public:
A(int a){ x=++a;}
~A(){cout<<x<<'\n';}
int get( ){return x;}
};
class B:public A{
int y;
public:
B(int b):A(b){y=get( )+b;}
B( ):A(5){ y=6; }
~B(){cout<<y<<'\n';}
};
void main(void)
    {
    B b( 5);
    }
```

8. 写出下面程序的输出结果。

```cpp
#include <iostream>
using namespace std;
 class B{
 public:
     void f1(){cout<<"B类中的函数 f1\n";}
     virtual void f2(){cout<<"B类中的函数 f2\n";}
 };
 class D: public B{
     void f1(){cout<<"D类中的函数 f1\n";}
     void f2(){cout<<"D类中的函数 f2\n";}
 };
 void main(void)
 {
     B a,*p;
     D b1;
     p=&a; p->f2();
     p=&b1;
     p->f1(); p->f2();
 }
```

9. 写出下面程序的输出结果。

```cpp
#include<iostream>
using namespace std;

class Base
{
public:
    void virtual f(){cout<<"Base::f()"<<endl;}
    void virtual g(){cout<<"Base::g()"<<endl;}
};
```

```
class A:public Base
{
public:
    void f(int a=0){cout<<"A::f()"<<endl;}
};
class B:public A
{
    void f(){cout<<"B::f()"<<endl;}
    void g(){cout<<"B::g()"<<endl;}
};
void main(void)
{
    B b;
    Base *p=&b;
    A *q=&b;
    p->f();   p->g();
    q->f();   q->g();
}
```

10. 写出下面程序的输出结果。

```
#include <iostream>
using namespace std;
class Complex{
public:
    Complex(){real=0;imag=0;}
    Complex(double r,double i){real=r;imag=i;}
    Complex operator+(Complex &c2);              //声明重载运算符的函数
    void display();
private:
    double real;
    double imag;
};
Complex Complex::operator+(Complex &c2){         //定义重载运算符的函数
    Complex c;
    c.real=real+c2.real;
    c.imag=imag+c2.imag;
    return c;
}

void Complex::display()
{ cout<<"("<<real<<","<<imag<<"i)"<<endl;}

int main(){
    Complex c1(3,4),c2(5,-10),c3;
    c3=c1+c2;                                    //运算符+用于复数运算
    cout<<"c1=";c1.display();
    cout<<"c2=";c2.display();
    cout<<"c1+c2=";c3.display();
```

```
        return 0;
    }
```

三、简答题（共 20 分）

1. 构造函数与析构函数在什么情况下被调用，派生类的构造函数需要对基类和派生类对象新增成员进行初始化，其执行顺序是什么？（5 分）
2. 结构 struct 和类 class 有什么异同？（5 分）
3. 简述成员函数、全局函数和友元函数的差别。（10 分）

四、编程题（共 20 分）

1. 下列 shape 类是一个表示形状的抽象类，area()为求图形面积的函数，total()则是一个通用的用以求不同形状的图形面积总和的函数。试从 shape 类派生圆形类(circle)、矩形类(rectangle)，并给出具体的求面积函数。（10 分）

```cpp
#include <iostream>
using namespace std;
class shape
{
    public:
    virtual float area()=0;
};
float total(shape *s[ ],int n)
{
    float sum=0.0;
    for(int i=0;i<n;i++)
     sum+=s[i]->area();
    return sum;
}
```

2. 编写字符串类 String 的构造函数、析构函数和赋值函数。（10 分）
已知类 String 的原型和 main 函数为：

```cpp
#include <iostream.h>
#include <string.h>
class String
{public:
String(const char *str=NULL);              //普通构造函数
String(const String &other);               //复制构造函数
~String();                                 //析构函数
String & operator=(const String &other);   //赋值函数
void show()
{cout<<m_data<<endl;}
private:
char *m_data;                              //用于保存字符串
};
void main()
{
    String str1("aa"),str2;
```

```
        str1.show();
        str2=str1;
        str2.show();
        String str3(str2);
        str3.show();
    }
```

模拟试卷 4

一、选择题(每题 2 分，共 20 分)

1. 已知程序中已经定义了函数 test，其原型是 int test(int, int, int);，则下列重载形式中正确的是(   )。

  A. char test(int,int,int);　　　　　　B. double test(int,int,double);
  C. int test(int,int,int=0);　　　　　　D. float test(int,int,float=3.5F);

2. 有如下程序：

```
#include<iostream>
using namespace std;
class Sample{
public:
    Sample(){}
    ~Sample(){cout<<'*';}
};
int main(){
Sample temp[2], *pTemp[2];
return 0;  }
```

执行这个程序输出星号(*)的个数为(   )。

  A. 1　　　　　　B. 2　　　　　　C. 3　　　　　　D. 4

3. 有如下程序：

```
#include<iostream>
using namespace std;
class A{
public:
    static int a;
    void init(){ a=1;}
    A(int a=2){init(); a++;}
};
int A::a=0;
A obj;
int main()
{
    cout<<obj.a;
    return 0;
}
```

运行时输出的结果是(   )。
  A. 0   B. 1   C. 2   D. 3

4. 在公有派生的情况下，派生类中定义的成员函数只能访问原基类的(   )。
  A. 公有成员和私有成员   B. 私有成员和保护成员
  C. 公有成员和保护成员   D. 私有成员、保护成员和公有成员

5. 有如下程序：

```
#include<iostream>
using namespace std;
class AA{
public:
    AA(){ cout<<'1'; }
};
class BB: public AA{
    int k;
public:
    BB():k(0){ cout<<'2'; }
    BB(int n):k(n){ cout<<'3';}
};
int main(){
    BB b(4), c;
    return 0;
}
```

运行时的输出结果是(   )。
  A. 1312   B. 132   C. 32   D. 1412

6. 建立一个有成员对象的派生类对象时，各构造函数体的执行次序为(   )。
  A. 派生类、成员对象类、基类   B. 成员对象类、基类、派生类
  C. 基类、成员对象类、派生类   D. 基类、派生类、成员对象类

7. 有如下程序：

```
#include <iostream>
using namespace std;
class Base{
public:
    Base(int x=0){cout<<x;}
};
class Derived : public Base{
public:
    Derived(int x=0){cout<<x;}
public:
    Base val;
};
int main(){
    Derived d(1);
    return 0;
}
```

程序的输出结果是（　　）。

    A. 0   B. 1   C. 01   D. 001

8. 在一个类体的下列声明中，正确的纯虚函数声明是（　　）。

    A. virtual void vf()=0;   B. void vf(int)=0;
    C. virtual int vf(int);   D. virtual void vf(int) { }

9. 有如下程序：

```cpp
#include <iostream>
using namespace std;
class B{
public:
    B(int xx):x(xx){++count; x+=10;}
    virtual void show()const
    { cout<<count<<'_'<<x<<endl;}
protected:
    static int count;
private:
    int x;
};
class D:public B{
public:
    D(int xx,int yy):B(xx),y(yy){++count; y+=100;}
    virtual void show()const
    {cout<<count<<'_'<<y<<endl;}
private:
    int y;
};
int B::count=0;
int main(){
    B *ptr=new D(10,20);
    ptr->show();
    delete ptr;
    return 0;
}
```

运行时的输出结果是（　　）。

    A. 1_120   B. 2_120   C. 1_20   D. 2_20

10. 打开文件时可单独或组合使用下列文件打开模式：
①ios_base::app；②ios_base::binary；③ios_base::in；④ios_base::out。
若要以二进制读方式打开一个文件，需使用的文件打开模式为（　　）。

    A. ①③   B. ①④   C. ②③   D. ②④

二、程序分析题（每小题4分，共40分）

1. 写出下面程序的输出结果。

```cpp
#include<iostream>
#include<iomanip>
```

```cpp
using namespace std;
//---------------------------------
class Date{
    int year, month, day;
public:
    void set(int y,int m,int d);
    void set(const string& s);
    bool isLeapYear();
    void print();
};//-------------------------------
void Date::set(int y,int m,int d){
    year=y; month=m; day=d;
}//--------------------------------
void Date::set(const string& s){
    year=atoi(s.substr(0,4).c_str());
    month=atoi(s.substr(5,2).c_str());
    day=atoi(s.substr(8,2).c_str());
}//--------------------------------
bool Date::isLeapYear(){
    return (year%4==0 && year%100!=0)||(year%400==0);
}//-------------------------------
void Date::print(){
    cout<<setfill('0');
    cout<<setw(4)<<year<<'-'<<setw(2)<<month<<'-'<<setw(2)<<day<<'\n';
    cout<<setfill(' ');
}//-------------------------------
Void main(){
    Date d, e;
    d.set(2000,12,6);
    e.set("2005-05-05");
    e.print();
    if(d.isLeapYear())
    d.print();
}
```

2. 写出下面程序的运行结果。

```cpp
#include<iostream>
using namespace std;
//-----------------------------------
class Student{
    int n;
    string name;
public:
    void set(string str){
        static int number = 0;
        name = str;
        n = ++number;
    }
```

```
        void print(){ cout<<name<<" -> students are "<<n<<" numbers\n"; }
};//----------------------------------
void fn(){
    Student s1;
    s1.set("Jenny");
    Student s2;
    s2.set("Randy");
    s1.print();
}//----------------------------------
Void main(){
    Student s;
    s.set("Smith");
    fn();
    s.print();
}
```

3. 写出下面程序执行结果。

```
#include<iostream>
using namespace std;
class MyClass{
public:
    MyClass(){cout<<"A";}
    MyClass(char c){cout<<c;}
    ~MyClass(){cout<<"XX";}
};
int main(){
    MyClass  p1,*p2;
    p2=new MyClass('B');
    delete p2;
    return 0;
}
```

4. 写出下面程序的输出结果。

```
#include<iostream>
using namespace std;
//------------------------------------
class StudentID{
    int value;
public:
    StudentID(){
        static int nextStudentID = 0;
        value = ++nextStudentID;
        cout<<"Assigning student id "<<value<<"\n";
    }
};//----------------------------------
class Student{
    string name;
    StudentID id;
```

```cpp
public:
    Student(string n = "noName"){
        cout <<"Constructing student " + n + "\n";
        name = n;
    }
};//------------------------------------
Void main(){
    Student s("Randy");
}
```

5. 写出下面程序的输出结果。

```cpp
#include<iostream>
using namespace std;
//------------------------------------
class A{
public:
    A(){ cout<<"A->"; }
};//------------------------------------
class B{
public:
    B(){ cout<<"B->"; }
};//------------------------------------
class C{
public:
    C(){ cout<<"C->"; }
};//------------------------------------
void func(){
    cout<<"\nfunc: ";
    A a;
    static B b;
    C c;
}//------------------------------------
Void main(){
    cout<<"main: ";
    for(int i=1; i<=2; ++i){
        for(int j=1; j<=2; ++j)
            if(i==2)C c; else A a;
        B b;
    }
    func(); func();
}
```

6. 写出下面程序的输出结果。

```cpp
#include<iostream>
using namespace std;
//------------------------------------
class Person{
    char* pName;
```

```cpp
public:
    Person(char* pN="noName"){
        cout<<"Constructing "<<pN<<"\n";
        pName = new char[strlen(pN)+1];
        if(pName)strcpy(pName,pN);
    }
    ~Person(){
        cout <<"Destructing "<<pName<<"\n";
        delete[] pName;
    }
};//---------------------------------
Void main(){
    Person p1("Randy");
    Person p2;
}
```

7. 写出下面程序的输出结果。

```cpp
#include <iostream>
using namespace std;
class A
{friend double count(A&);
public:
A(double t, double r):total(t),rate(r){}
private:
    double total;
double rate;
};
double count(A& a)
{
a.total+=a.rate*a.total;
return a.total;
}
int main(void)
{
    A x(80,0.5),y(100,0.2);
    cout<<count(x)<<','<<count(y)<<'\n';
    cout<<count(x)<<'\n';
    return 0;
}
```

8. 写出下面程序的输出结果。

```cpp
#include<iostream>
using namespace std;
class A{
    int x;
public:
    A(int x=0)
    { this->x=x;}
```

```cpp
        virtual void f(){cout<<x<<endl;}
};
class B:public A{
    int y;
public:
    B(int x,int y=1):A(x)
    {
        this->y=y;
    }
    void f(int a){cout<<y<<endl;}
};
void main(void)
{
    A a1(10),*pa;
    B b1(20,30);
    a1.f();
    pa=&a1;
    pa->f();
    pa=&b1;
    pa->f();
}
```

9. 写出下面程序的运行结果。

```cpp
#include <iostream>
using namespace std;
class Complex{
public:
    Complex(){real=0;imag=0;}
    Complex(double r,double i){real=r;imag=i;}
    Complex operator+(Complex &c2);        //声明重载运算符的函数
    void display();
private:
    double real;
    double imag;
};
Complex Complex::operator+(Complex &c2){   //定义重载运算符的函数
Complex c;
    c.real=real+c2.real;
    c.imag=imag+c2.imag;
    return c;
}

void Complex::display()
{ cout<<"("<<real<<","<<imag<<"i)"<<endl;}

int main(){
    Complex c1(3,4),c2(5,-10),c3;
    c3=c1+c2;                              //运算符+用于复数运算
```

```
        cout<<"c1=";c1.display();
        cout<<"c2=";c2.display();
        cout<<"c1+c2=";c3.display();
        return 0;
    }
```

10. 写出下面程序的运行结果。

```
#include<iostream>
#include<cmath>
using namespace std;
//------------------------------------
class Shape{
protected:
    double xCoord, yCoord;
public:
    Shape(double x, double y): xCoord(x),yCoord(y){}
    virtual double area()const{ return 0.0; }
};//----------------------------------
class Circle : public Shape{
protected:
    double radius;
public:
    Circle(double x, double y, double r): Shape(x,y),radius(r){}
    double area()const{ return 3.14 * radius * radius; }
};//----------------------------------
class Rectangle : public Shape{
protected:
    double x2Coord, y2Coord;
public:
    Rectangle(double x1, double y1, double x2, double y2)
      : Shape(x1,y1), x2Coord(x2), y2Coord(y2){}
    double area()const;
};//----------------------------------
double Rectangle::area()const{
    return abs((xCoord-x2Coord)*(yCoord-y2Coord));
}//-----------------------------------
void fun(const Shape& sp){
    cout<<sp.area()<<"\n";
}//-----------------------------------
Void main(){
    fun(Circle(2, 5, 4));
    fun(Rectangle(2, 4, 1, 2));
}
```

三、简答题(共 20 分)

1. 若程序员没有定义复制构造函数，则编译器自动生成一个缺省的复制构造函数，它可能会产生什么问题？(5 分)

2．重载函数是根据什么来区分？（5 分）

3．简述结构化的程序设计、面向对象的程序设计的基本思想。（10 分）

四、编程题（共 20 分）

1．设计一个"学生"类，其中"学生"类具有"姓名"、"学号"、"性别"三个数据成员，以及构造函数，复制构造函数和用于显示学生信息的成员函数。要求写出完整的类定义，包括成员数据和成员函数的定义。（10 分）

2．编写一个程序，定义一个抽象类 Shape，由它派生 3 个类：Square（正方形）、Trapezoid（梯形）和 Triangle 三角形。用虚函数分别计算几种图形面积并求它们的和。要求用基类指针数组，使它每一个元素指向一个派生类对象。（10 分）

类 Shape 和函数 main() 如下：

```cpp
class Shape
{
public:
    virtual double area()const=0;
};
void main()
{
    Shape *p[5];
    Square se(5);
    Trapezoid td(2,5,4);
    Triangle te(5,8);
    p[0]=&se;
    p[1]=&td;
    p[2]=&te;
    double da=0;
    for(int i=0;i<3;i++)
    {
        da+=p[i]->area();
    }
    cout<<"总面积是："<<da<<endl;
}
```